SUNDIALS

HISTORY, THEORY, AND PRACTICE

René R. J. Rohr

WITH A FOREWORD BY Henri Michel
TRANSLATED BY Gabriel Godin

DOVER PUBLICATIONS, INC.
New York

Published in Canada by General Publishing Company, Ltd., 30 Lesmill Road, Don Mills, Toronto, Ontario.

Published in the United Kingdom by Constable and Company, Ltd., 3 The Lanchesters, 162–164 Fulham Palace Road, London W6 9ER.

Bibliographical Note

This Dover edition, first published in 1996, is an unabridged and slightly altered republication of the first English translation (published by University of Toronto Press, Toronto, in 1970) of the original French work published by Gauthier-Villars, Montrouge, France, in 1965 under the title *Les Cadrans solaires*. For the Dover edition the plates on page 2, originally in color, appear in black and white in the text and in color on the inside back cover. Also, the original color title-page plate has been moved to the inside back cover.

Library of Congress Cataloging-in-Publication Data

Rohr, René R. J.

    [Cadrans solaires. English]

    Sundials : history, theory, and practice / René R. J. Rohr ; translated by Gabriel Godin ; with a foreword by Henri Michel.

       p.   cm.

    Originally published: Toronto : University of Toronto Press, 1970.

    Includes bibliographical references and index.

    ISBN 0-486-29139-1 (pbk.)

    1. Sundials.   I. Title.

QB215.R613   1996

681.1′11—dc20

                                          95-45661

                                          CIP

Manufactured in the United States of America

Dover Publications, Inc., 31 East 2nd Street, Mineola, N.Y. 11501

Contents

Foreword

As a specialist, I have read most of the books about sundials and yet I must sincerely recommend this one and congratulate the author on it.

Formerly a ship captain, René R. J. Rohr had already revealed himself as a fine observer and as a person extremely sensitive to the beauty of things in his books on travel. He is therefore well equipped to pass judgment on an instrument that is simultaneously scientific, historical, and artistic. His lively pen, his direct and simple style, his expressive vocabulary, his avoidance of pedantry, his conciseness in the exposition of his thoughts make his book a pleasure to read.

Too many writers, in dealing with this topic, would have accumulated the references or exhibited their erudition, or even taken a detached attitude of superiority. René R. J. Rohr knows that true humility consists in ignoring one's own humbleness and never taking pride in it. He thinks first about his reader, of what will interest him. He does not try to teach: he only wants us to develop understanding.

Once we have read these lively, picturesque, and clearly written pages, once we have gained a precise idea of this gnomonics which has attracted so many mathematicians, we will feel much sympathy for the author. Indeed he has fulfilled the three conditions required for a good book: have something to say; say it; don't say anything else.

Henri Michel, International Academy of the History of Science

Preface

Today, in the middle of the twentieth century, the sundial still enjoys a popularity which it would be futile to attribute to any economic advantage; it has not sunk into oblivion or been deposited for ever in the museums along with the staff of Jacob or the astrolabe. Still, the time has long passed since the clockmaker needed it to check his watches, consulting every day the "table of equation" carefully kept under glass.

It is not even because a well-designed dial gives the time with more precision than many of our watches that dials have reappeared on the walls of houses or in the greenery of gardens. The real reasons are more fundamental. We must look elsewhere.

Years ago, when we saw some venerable and ancient dial on the façade of a church or on old country houses, we were fully conscious of being in the presence of a witness to a past in which time moved more slowly, where life was not so complicated, and when it was easier to relax and meditate a little. Nowadays the watches and the clocks give the time with the accuracy necessary for the hectic pace of industry and the working hours. Absorbed by the grind of modern life, man is carried against his own will into its vortex, but the need to resist the threat to his personality and his peace of mind soon asserts itself. He is drawn towards older things, and if he discovers among the ivy of an old wall a sundial un-

disturbed and unspoiled by the decay of time or the wear of the years, he will feel himself steeped in an atmosphere of peace, concentration, and rest. A kind of eternal magic radiates from each sundial, awakening obscure recollections of a past that has now become a vestigial subconscious, reminding man of the time when his life was clearly felt to be part of the life of a powerfully animist universe. Wherever there is a sundial, it will attract our attention and its contemplation will recall the homage rendered by our remote ancestors to the sun, the first among all the "gods" to be worshipped.

A dial located in the middle of a garden is a decorative object that graces all that surrounds it. It fits beautifully among plants and flowers. The rose bush is its ideal neighbour; life for the sober dial on its moss-covered pedestal or the shimmering rose bush depends on light. But one lasts long and measures the external; the other quickly fades, and symbolizes the quick passage of life and death. And, on the other hand, a bare and ugly wall, carrying a simple dial, can also become alive and interesting. The modern architect knows this very well and uses the sundial to re-establish the equilibrium on an uneven façade, to fill a hole, or simply to improve the appearance of a house.

The role which the sundial plays as an intermediary between

more accurate results. Here naturally, the reader who is familiar with the procedures of descriptive geometry will have no difficulty. Our intention is to give everybody the information needed for designing dials. For this reason elementary details are elaborated and there is some repetition in order to present the problem under slightly different aspects. (In the case of the analemmatic dial, we must admit that we started with the *épure* since no direct and practical method exists for its construction. Because of its intrinsic interest and in spite of its rarity, we have studied it in such a way that its construction with the *épure* is possible under any circumstances.)

When it comes to designing the dial by the sole means of mathematics, we had in mind those of our readers with some knowledge of trigonometry or its rudiments. For the sake of simplicity, however, we have given in chapter four the formulae which are applicable to the classical dials but which are almost never to be found in the standard treatises. In this way, whenever we studied a particular dial, we could definitively relate it to the appropriate form. Nevertheless, for the sake of interest, we have given an elementary example of this type of mathematical calculation for the horizontal dial, which can serve as a guide to interested readers for a study of the other dials. Besides its value as a check, the mathematical approach is very useful in the design of the analemmatic dial for which the *épure* is laborious, whereas the analytical approach requires only a few minutes of work.

The author had recourse to many persons and many institutions in collecting all the documents relevant to the writing and the illustration of the present book. He cannot name them all but feels that he must mention Professor Harro Heinz Kühnelt of the University of Innsbruck; Mr A. M. Burg of the Museum of Haguenau; Professor G. C. Shephard of the University of Birmingham; Mr Ed. Remouchamps of the Musée de la Vie Wallonne in Liege;

the apparent path of the sun and our life makes it a first-class pedagogical tool. The joy of making one is followed by the equal joy of exploring the fascinating and easily acquired knowledge of the motion of our earth. This is truly learning by discovery! The sundial carrying its daily arcs retraces every year the complete trajectory of the earth around the sun.

These three factors – its link with the sun and with all the sun means to us, its beauty to the eye and to the reflective mind, and its value as an instrument of scientific understanding – are such that the old tag *tempus edax rerum* ("time gobbles up everything") will not apply to the sundial as long as man feels and thinks. But the principle of the sundial has been forgotten by many people, and the ugly name used for its theory – gnomonics – helps also in giving rise to a vulgar misconception that a sundial is almost some kind of cabalistic object.

But the theory is quite simple, and in this book we attempt to make it available to everybody. We first acquaint the reader with the history of the sundial – a task for which the research brought us much pleasure.

The mechanical construction of a sundial without any knowledge of its theory can bring to its maker, however, no satisfaction to the rational part of the mind, so it seemed appropriate to us to discuss in another introductory chapter the basic elements of cosmology in so far as they are related to the motion of the sun, the moon, and the earth. The knowledge of these facts, of interest in themselves, forms the base for any useful application of gnomonics.

In the technical part of the book, the explanations are designed in succession for various types of readers. We start with the description of a practical method of construction for which no special knowledge is required. Next we deal with the *épure*★ which gives

★ A full-scale working drawing, usually traced on a wall or floor.

Mr A. G. Coromilas, Athens; Mrs Anny R. Robinson of Phillips Academy, Andover, USA; Mr Berg of the Botanical Garden of Bremen; Dr Rodolfo Capelli and Miss Maria Luisa Bonelli of Florence; Mr H. Reinhard of the Historical Museum of Basle; Mr Victor Beyer of the Museum of Strasbourg; Mr Pierre Noël, Paris; Miss Renée Rohr, Strasbourg; Mr Desmet, Grenoble; Mr Pierre Chaumeton, Marseille; the Municipality of Annecy; the Museum of Bourg-en-Bresse; and the Gübelin Company of Lucerne.

We wish to make special mention of Mr Carl Moureau, the man behind the Maison du Cadran Solaire of Carcassonne, one of the rare firms in France dealing with the installation of dials. Mr Moureau has made his rich documentation available to us and has aided us in all our enquiries.

The author is particularly grateful to Mr Henri Michel who has been kind enough to write the Foreword to this book and so to lend it the prestige of his great international authority in the very specialized field of the history of astronomical instruments.

René R. J. Rohr, Muhlbach-sur-Bruche, France

SUNDIALS

HISTORY, THEORY, AND PRACTICE

B

A

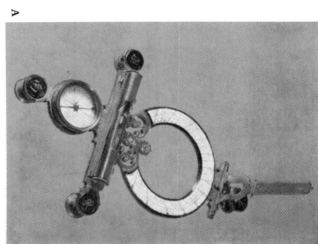

C

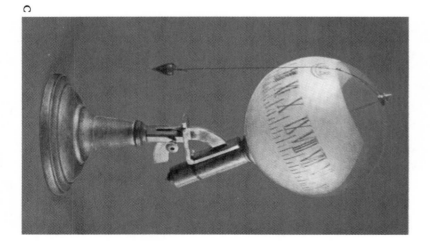

PLATE A   Horizontal sundial made in Canton by the East India Company around 1800 (CNAM; *photograph: Rapho Ciccione*)

PLATE B   Azimuth or magnetic dial by Hermand, eighteenth century (CNAM; *photograph: Rapho Ciccione*)

PLATE C   Equatorial sundial, also called solar clock. Rimbaud, second half of the nineteenth century (CNAM; *photograph: Rapho Ciccione*)

PLATE D   Equatorial sundial, an imitation of the scaphion of the Ancients, Deslincourt (CNAM; *photograph: Rapho Ciccione*) (See inside back cover)

CHAPTER ONE

History of the Sundial

In Europe the beginning of the great human adventure took place amid the frosty surroundings of an ice age. One may try to imagine virgin and raw landscapes, framed by a sky more often gray than sunny, but always bathed in the fresh cool air of early morning; here and there small groups of squat men are seen wandering, men whose primitive looks are suggested to us from Heidelberg to Cromagnon by a meagre array of fossil remains.

From dawn, till dusk these remote ancestors of ours had to fight endlessly for survival, against hunger, cold, and the various perils which a merciless nature put in their path. From dawn till dusk, indeed, because it was the only time reckoning they knew. The rhythm of day and night which had regulated all life on earth for a billion years from the life of the algae to that of the trees and the powerful mammoth was also the rhythm of their life. They have left us nothing to witness this fact, but we know it could not have been otherwise. For thousands of years, the sunrise and the sunset were the only signals for a change in their activity. A remote echo of this elementary state of things reverberates through the Bible where the book of Genesis (1:5) brings us in its first verses the lapidary sentence:

God called the light Day and the darkness he called Night. And there was evening and there was morning, one day.

Painfully thousands of years went by and man did not perish. Indeed he flourished and suddenly he has in his hand a tool – a poor tool, a stone, a raw or roughly hewn piece of flint. But the man who held it did not know that he as well as the whole of humanity had just taken a decisive step. There were many more steps before the great day came when settled life was established.

Gradually but obscurely, man faced a new need. Every primitive being fears the hidden dangers of darkness. And each time a man had to leave his tent of skins or the huts of his tribe to go to some remote place, he had to be able to determine when he should turn back so that he would not be caught on the way by sunset. Undoubtedly, since he spent his life outside, man developed the habit of watching the daily course of the sun and he had probably learned that he could move away from the group as long as the sun rose but that he had to be on his guard from the moment it started going down. His experience must have followed the progress of his tools. He realized that the shadow of the trees shortened in the first part of each day and that it lengthened later. Certainly the instant of transition between these two movements must have soon played an important role in the slow but progressive organization of his life. It was the first step, hesitant as yet, towards the clock-dominated and hectic life of his remote descendants in the twentieth century.

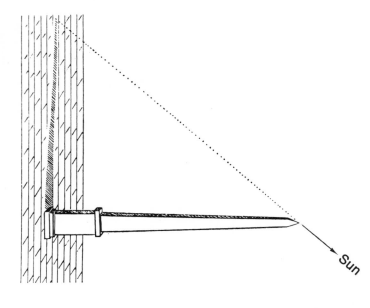

FIGURE 1   The gnomon – here an obelisk. Its shadow has the same length twice in one day: once in the morning, once in the evening

But can we not go a little further and imagine that primitive man had learned to exploit the changes in the length of his own shadow? Or rather that, as his intelligence sharpened, he had the time, through the years, to think of using a rod of a given length, the shadow of which, when set vertically, could help fix the moment when he should meet a companion equipped, perhaps, with the same instrument.

This instrument would have been the first gnomon – the principle of which was used centuries later, when the peoples around the shores of the Mediterranean erected the first stone obelisks on the great squares of their cities. Museums overflow with all kinds of implements from the Neolithic age: thousands of flaked flints have been found, but a simple little rod, besides being highly perishable, would probably be too plain an object to have attracted attention.

All this, admittedly, is pure conjecture. But, happily, progress is not simultaneous everywhere. Do we not have to help us in this kind of research among the reserves of primitive men in the twentieth century in Australia, Melanesia, and New Guinea, many samples of tribes which have not yet gone beyond the Neolithic age? And observation of their customs tends to support the hypothesis just proposed.

In fact, if the search for positive clues on the measure of time turns out to be so difficult for the early stages of human civilization, the archaeologists who have studied the same problem for the initial eras of history are in no better position. There is no doubt that the gnomon was the first instrument used for the measurement of time by various peoples and that they used the length of the shadow and not its direction in order to do this. Gnomon is a Greek work meaning "pointer." In contrast with our mechanical watches, it could not be used to delineate an interval or lapse of time, but indicated rather a given moment (figure 1).

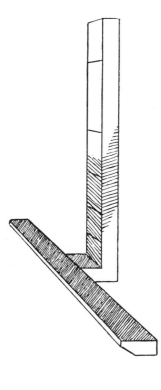

FIGURE 2   Egyptian dial of the time of Thutmosis III (fifteenth century BC). The most ancient dial known

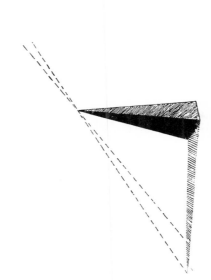

FIGURE 3   The apparent diameter of the sun has been grossly exaggerated in order to show the decrease in sharpness of the edge of the shadow with the increasing height of the gnomon

The intrinsic simplicity of any apparatus used for this purpose makes its identification among other objects found in an excavation rather difficult. Besides, scholars have been faced with an even more elusive problem since the finds are scarce and thinly spread geographically. They are thus compelled either to rely on written documents or to try to interpret as correctly as possible the purpose of the implements discovered when this purpose seems to be related to their special field of study.

Groping through history with this Ariadne's thread, we learn from the papyri that by about 1450 BC gnomons in the form of obelisks were used in Egypt for the measurement of time and the setting up of calendars. Even earlier, Thutmosis III (1501 to 1448 BC) had carried with him on his trips a portable dial, the nature of which is not known to us; it might have resembled an object found and dated as belonging to the same era and which is the oldest sundial known. This object does not look at all like a gnomon, although it is constructed on the same principle. It consists of a piece of L-shaped stone about 30 cm long, supporting on its small end a second stone of the same length, but straight and perpendicular to it. The whole thing was oriented on a horizontal plane as indicated in figure 2, i.e., the long stem of the L-shaped stone was placed opposite the sun. This stem was covered by the shadow of the cross piece and on its surface were etched divisions which indicated the hours, according to the height of the sun. The duration of these hours could not be constant from one day to the next because of the variation of the sun's declination during the course of a year.

The chronological Ariadne's thread which we have proposed to follow brings us from Egypt to China, a thousand years before Christ. According to old documents from this country, the gnomon was commonly used as an instrument for astronomical observations. Not only had the Chinese already succeeded in locating the astronomical meridian with its help, but they had also succeeded in fixing the dates of the solstices and had even calculated the inclination of the ecliptic on the plane of the equator. The value of this inclination is 23° 27′ in round numbers; the result obtained in China was 23° 54′, a very respectable estimate considering the means used to obtain it. It goes without saying that such work cannot be improvised, so we are justified in thinking that gnomons must have been in use in China from very early times. Let us stress that what we have been discussing here are the first appearances of the gnomon in written records. These records also tell us that astronomical observations were initiated in China in the era of Yao, an emperor shrouded in legend, who lived in the twenty-third century before Christ; it is said that he had two of his astronomers executed because they failed to predict an eclipse of the sun.

The same sources also reveal that the perforated gnomon was known in China from earliest times. The peoples living on the shores of the Mediterranean discovered this instrument and its use much later. Its invention was there attributed to the Arab astronomer Ibn Junis who lived at the end of the tenth century AD.

As the gnomon becomes higher, the determination of time by this means becomes more accurate. But since the sun is not a point but a disc with a certain apparent diameter, the edge of this shadow lacked definition because of the penumbra which surrounded it (figure 3). To overcome this difficulty, the Chinese installed a circular disc, through which a round hole was pierced, on top of the gnomon. The shadow then left on the ground formed a little round spot, the image of the sun, the centre of which could be easily determined (figure 4). Finally, around 500 BC a uniform height for all the Chinese gnomons was prescribed by law under the threat of severe penalties.

The absence of written documents among the other nations

does not prohibit the possibility that the gnomon was also known there. Well into this century, for example, we were unaware of the old Egyptian dial of the fifteenth century BC mentioned above.

Among the Hindus, one finds from early times gnomons surrounded by concentric circles which made it easier to determine the true south, i.e., the meridian. We shall return to this technique, for it is still quite widely used. This gnomon was rapidly adopted by other peoples of the period and the term "Hindu circles," by which it is currently designated, leaves no doubt of its origin.

In Mesopotamia, the Babylonians and then the Chaldeans enjoyed a high reputation as astronomers. Herodotus describes in his *History* (fifth century BC) the annual allotment of lands in Egypt and the knowledge of geometry derived from it – a knowledge which the Greeks imported into their own country. He says, "the Greeks learned from the Babylonians the use of the polos, of the gnomon, and of the division of the day into twelve parts" (II, 109). In the very rare references to sundials in the texts of that era, the word *polos* appears here for the first time. Herodotus seems to

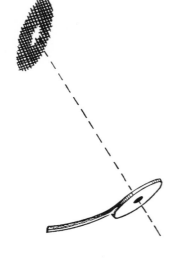

FIGURE 4   The pierced gnomon throws an elliptic spot of shadow with a bright ellipse inside it whose centre it is easy to determine

make a distinction between the *polos* and the gnomon. Over the centuries the word *polos* has entered the language of the gnomonists of some countries to designate dials in which the direction of the style is parallel to the axis of rotation of the earth or again to designate the style itself when it is oriented in this fashion rather than along the vertical as is the case of the gnomon. There is a big difference here. Indeed, as we shall see, because of the sun's declination, none of the ancient gnomons discussed so far could give constant results throughout the seasons of the year. The length of the shadow at a given hour of one day does not correspond to that at the same hour the next day. There will be little enough difference between them at the solstices, but it will be considerable during the equinoxes; between a summer day and a winter day there is no possible comparison. On the *polos* dials, however, with the style parallel to the axis of the world, the reading of the hours is done with the same precision throughout the seasons. So we may assume that what Herodotus calls a *polos* is a dial which could have existed in Egypt in his time but about which no information has survived.

We must remember in connection with Herodotus' attribution of sundials to the Babylonians that he was disinclined to check on all the items of information supplied to him during the course of his travels. Taking account of the high level of Greek culture in the days of Herodotus and of the immense contribution of Greek thought in the domain of mathematics, we may posit that, in spite of the rarity of the hints in the written documents, the gnomon had been known to the Greeks for some time. In any event, in 560 BC, and therefore during the lifetime of Herodotus, Anaximander of Miletus installed a sundial in Lacedaemonia, the nature of which is not known to us. We can assume that, like the Chinese, the Greeks were using the gnomon as an instrument of observation around 600 BC.

We know too that in Herodotus' lifetime small portable dials of a peculiar shape were used in Egypt: the shadow of one edge fell on a horizontal plane divided into hours or on the steps of a small stairway, or at times on both of them at the same time. In this last instance, their shape suggests certain forms of Egyptian architecture.

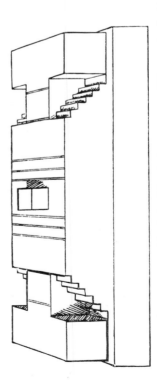

FIGURE 5 Egyptian portable dial of the sixth century BC. We note its decidedly Mesopotamian appearance, given to it by the stair steps. It is not impossible that King Ahaz' altar mentioned in the Bible in IV Kings had this appearance. FIGURE 6 A portable Egyptian dial of the fourth century BC. The time is read in the column of the month between the converging lines. If the east is to the left of the figure and if it is the month of May, we have just passed the first third of the third hour of the day

ture or even, on account of their steps, the ziggurats of the old cities of Mesopotamia (figure 5). (We shall revert shortly to the strong probability of a Chaldean origin for these objects.) A later sample of this type of dial, from the fourth century BC, exhibits a definite improvement over the preceding type. The shadow fell on an inclined plane, on which lines scaled in hours had been drawn along the line of the steepest slope in such a way that the scale on each line corresponded to a given month of the year. Figure 6 shows a schematized representation of this dial, the first one in the history of gnomonics to take account of solar declination (the changing height of the sun at midday over the year). These later Egyptian dials imply, as does the earlier one in figure 2, that the shadow edge had been oriented first toward the sun.

It should be stressed that one thousand years had elapsed between the dial shown in figure 2 and those of figures 5 and 6, and that during that time sundials had obviously been in constant use. But we know absolutely nothing about them because of the scarcity of gnomonic finds in Egypt.

The presence of stair steps in the dial of figure 5 may serve to clarify some puzzling statements in the Bible about a sundial belonging to a certain king Ahaz, who reigned in Judea from 740 to 728, two centuries before the known appearance of the step dial in Egypt:

And Isaiah said "This is the sign to you from the Lord, that the Lord will do the thing that he has promised: shall the shadow go forward ten steps, or go back ten steps?

And Hezekiah answered, "It is an easy thing for the shadow to lengthen ten steps: rather let the shadow go back ten steps."

And Isaiah the prophet cried to the Lord: and he brought the shadow back ten steps, by which the sun had declined on the dial of Ahaz. (Kings 20:9–11)

Elsewhere, we read (Isa. 38:8):

"... Behold, I will make the shadow cast by the declining sun on the dial of Ahaz turn back ten steps." So the sun turned back on the dial the ten steps by which it had declined.

In his days the sun went backward, and he lengthened the king's life.

In the book of Sirach or Ecclesiasticus (in the Apocrypha) we find another allusion (Ecclus. 48:24) to this famous dial of Ahaz:

The one thing we learn with certainty from these quotations is that the dial of Ahaz was provided with steps. But a close reading of the Book of Kings allows us to draw some additional inferences. Indeed, at a time when Ahaz was at war against both the Arameans and the Israelites, he thought it would be proper to call to his help Tiglath-pileser, the then all-powerful king of Assyria. It happened that the latter, also engaged in a war campaign, had occupied the city of Damascus for almost ten years. The two kings agreed to meet in this city and, when Ahaz arrived, he was brought to an altar which interested him to such a point that, according to II Kings 16:10, 11:

When King Ahaz went to Damascus to meet Tiglath-pileser king of Assyria, he saw the altar that was at Damascus. And King Ahaz sent to Uriah the priest a model of the altar, and its pattern, exact in all its details.

And Uriah the priest built the altar in accordance with all that King Ahaz had sent from Damascus, so Uriah the priest made it before King Ahaz arrived from Damascus.

There are two possibilities. Either this extraordinary Damascus altar was already an ancient marvel – let us say, from before Ahaz's war, at least – which was certainly known to the great men of the surrounding countries, and therefore to Ahaz and his priests, because even at that time, the distance between Damascus and Jerusalem was far from prohibitive. Or this altar was a novelty built by the Assyrians after their arrival in the country in honour of their own gods. In the latter instance we would like to believe, taking special account of the enthusiasm of King Ahaz, that it was the step sundial he had built in Jerusalem his capital, and that this construction was of Chaldean origin since the Chaldeans, the then masters of gnomonics, were the vassals of Tiglath-pileser. Isaiah's hostility towards the new altar and the accusations which the Bible throws at Ahaz tend to prove the foreign and heretical origin of the whole installation. For if it were Chaldean, we would be able to trace, with some certainty, the step-type Egyptian sundial from the shores of the Nile back to Chaldea.

The reader will forgive our stopping for a rather long time by this curious dial of Ahaz, but he should note that no previous dial had such good coverage in the old documents and that no other, let us hasten to add, would give rise thereafter to such a complex and abundant literature. Josephus Flavius, the Jewish historian of the beginning of the Christian era started it; it has been pursued in the studies of Pedro Nunez (1564), Camille Flammarion (1885), and Claudio Pasini who deals with it in his *Orologi solari* of 1900, to mention only a few. Scores of books have been written about what is called the miracle of Ahaz' sundial. Since we really do not know the nature of this sundial, some authors have tried to picture a dial whose functioning would involve the backward motion of the shadow. To quote one example from thousands, we find in the Museum of the Philosophical Society of Philadelphia a sixteenth-century sundial in the form of a bowl provided with a gnomon, the work of the celebrated craftsman, Christopher Schissler of Augsburg. When water is poured in the bowl, refraction of the light rays causes the shadow of the gnomon to recede.

The problem involving the retrograde motion of the shadow on Ahaz' sundial has never been solved; the Bible maintains it is a miracle. The work of so many scholars has, nevertheless, led to a

clarification of the Hebrew text of our first quotation from the book of Kings, where the word *ma'aloth* is used three times; *ma'aloth* may mean either line, degree, or the step of a staircase. In translating this word into Latin, St Jerome, the author of the Vulgate, had the unfortunate idea of using a different equivalent each time, with the result that his version has given rise in the course of the centuries to the most inconceivable confusion and the most absurd hypotheses.

At the time of Ahaz, the Chaldeans had grown into the habit of dividing that portion of the heavens containing the orbits of the sun and the planets into twelve parts. They gave to each of these parts the name of a constellation. Their sum comprised the zodiac. The day, that is to say, the interval between sunrise and sunset, was divided into twelve hours and was therefore of variable duration. The Chaldean priests predicted the eclipses of the sun and of the moon.

In the third century BC, one of these priests, Berossos by name, designed a dial in Egypt in the shape of a hemisphere hollowed out of a rectangular block of stone and with a gnomon set upright in the centre of the hollow so that its tip occupied the centre of homothety★ between the celestial vault and the hemisphere. The hemisphere was the image of the other, and each point in it corresponded with a point of the celestial vault. To each arc of a circle traced daily by the sun there corresponded an arc of a circle in the hemisphere, which was its exact image. This was a brilliant idea. Berossos drew inside his dial the arcs corresponding to the position of the sun at the equinox and the solstices; in accordance with the Chaldean usage of dividing the day into twelve hours, he divided the resulting zone into twelve sectors (figure 7).

The dial indicated hours which varied according to the length

★ Similar orientation, usually of geometric figures.

of the day for the given solar declination, i.e., for the various seasons. Later called the *hemispherium*, it was a portable dial constructed to be as light as possible. Since the entire portion of the hemispherium below the image of the solstice of Cancer was unnecessary, Berossos designed a new model similar to the first but with this portion removed and the vertical gnomon, which could not now be put at the hemispherium, replaced by a horizontal gnomon (figure 8). To distinguish this new model from the hemispherium it was called a *hemicyclium*.

These dials spread rapidly in varying forms throughout the ancient world. In Greece they were called *heliotropes*, and Aristarchus of Samos, around 250 BC, created a variety called *scaphions*, which in gnomonic nomenclature later designated all hollow dials. According to a text of the Arabic mathematician Al Battani (Albategnius), the Arabs were still using this type of dial in the

tenth century AD.

Some rare dials of a slightly different type have been found in Italy, Greece, and Egypt: the hemisphere was replaced by a cone-shaped excavation but the lines were traced on the same principle. The result was evidently the same. Perhaps this newer form was easier to read, but mathematically it lacked justification. It is thought that the appearance of these strange variants – called conical dials – followed the publication of the famous treatise on conic sections by Apollonius of Pergae in the second half of the third century BC. It could have been an offering to the god of novelty, the current fashion. But it should be noted, in fairness to the Greek artisans, that these dials were easier to make.

We do not know why, in Rome at least, dials of the scaphion type never achieved a reputation for being accurate; the reduced dimensions of the instruments or inattention on the part of the

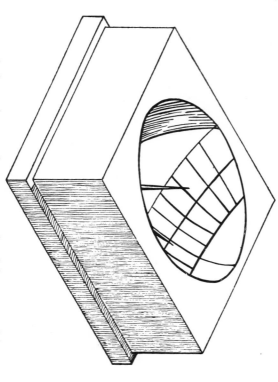

FIGURE 7 The hemispherium of Berossos reproduced the passage of the sun on the celestial vault in a reduced scale

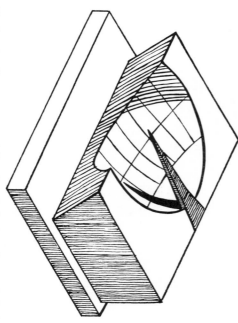

FIGURE 8   The hemicyclium was essentially a hemispherium from which the useless portion located to the south of the image of the Tropic of Cancer was removed. The instrument was thus lightened. It would have been possible to lighten it further by removing the portion to the north of the image of the Tropic of Capricorn

makers to their proper construction and installation may have been responsible. The transition from a hemisphere to a cone helped make reading easier but it could not in any way compensate for the negligence of the makers. Seneca, speaking about this inconvenience, said that philosophers could agree more easily among themselves than do the dials.

In Greek and Roman times the day was still divided between sunrise and sunset into twelve hours. Varying with the seasons, these hours were usually called the "temporary" hours.★ The interval between sunset and sunrise was also divided into twelve equal hours so that during the day and the night time was measured in hours that differed in duration, except during the equinox. To measure the nocturnal hours and those of the sunless days, the clepsydra, the water clock, an old Mesopotamian invention of the sixth century BC, was used; it was based on the principle that a given amount of water always takes the same time to fall drop by drop from one container to another. None of these delicate instruments seem to have survived the ravages of time, but their principle is known to us from a description in Vitruvius' *De architectura*.

Since the equinoctial days were the only ones where the hours had the same duration during the day and during the night and for all the latitudes, they became the standard for indicating time with precision and for astronomical data. These equinoctial hours corresponded to the twenty-four hours of our day; the "temporary" hours in the latitude of Rome corresponded roughly to one and a quarter of our hours in summer and to three-quarters of these same hours in winter.

There must have been a time when the use of the scaphion, in one form or another, was general throughout the ancient Mediterranean world, or at least in the cities. We have seen that Aristarchus devised a few of them but gradually the science of their construc-

★ *Horae temporariae* – hours measured according to sun time and consequently variable.

tion spread so widely that there were specialized shops in Greece and Italy where they were built and where an allowance was made for the latitude in which they would be used. Some made of travertine marble have been found in Pompeii, in Castelnuovo, and in Rignano. The hour lines of these dials carried no label; one had then to count them. A hemicyclium, labelled with Greek letters, was discovered in 1852 at the foot of Cleopatra's Needles when these two famous obelisks were still in Alexandria. The Greek word for life could be made out of the first four hours of the afternoon. Some people wish to see in this possibly fortuitous arrangement of letters a first example of the use of a motto. The Vienna Museum preserves a beautiful hemicyclium originating from the Siebengebirge, close to the shores of the Rhine, the hollow of which has a diameter of thirty centimetres and whose sides are decorated with reliefs of vases.

Rome got her first sundial in 293 BC, a scaphion located close to the temple of Jupiter Quirinus. About thirty years later, a dial looted from Catana during the First Punic War was brought to a public square. Catana was of Greek origin and the dial, most likely Greek as well, must have attracted the attention of the Romans. Since it had been devised for the latitude of Sicily, it could not record the hours accurately in Rome. The Romans, with their usual indifference to the exact sciences, did not discover the error until a century had passed. (Pliny reports this incident with bitter irony.) It was Marcus Philippus, whose duty as a censor included the watching of public dials, who realized the significance of a properly designed dial. The presence of these dials did not exclude the co-existence of gnomons. In the Field of Mars, as before, the great obelisk of Augustus continued to inform the passers-by. In old Rome, old habits died hard, and we shall see that the gnomons survived there for many centuries.

Nevertheless, the dials must have multiplied rapidly because we read two centuries later in the *Boeotian* (probably by Plautus) the following explosion of a Roman citizen's ill humour:

Let the gods damn the first man who invented the hours, the first man who set up a sundial in this city! For our misfortune he has chopped up the day into slices.

When I was young, there was no other clock but my belly. For me it was the best and the most accurate clock; at its call we ate, unless there was nothing to eat. Now, even if there is an abundance, we have to eat only when it pleases the sun. The city is full of sundials, but we see almost everybody crawl around half dead with hunger.

To go back to Greece, in the *Assembly of Women* by Aristophanes, the woman Praxagora says to her husband: "When the stoicheion is ten times as long as your foot, all you will have to do is perfume yourself and come to dinner" (5: 651–652). In this instance the stoicheion was the length of the husband's shadow.

Excerpts of this type gleaned from the comedies almost always refer to meal hours. When it came to the gnomon, the authors never missed a chance to exploit the possibility of confusion between the morning and evening hours. The tone stops being light and it is no longer a scene in a comedy when the Greek Philoxene is advised by his doctor: "If you have any business left to settle, go ahead, do not lose a minute, because you will die within the space of seven feet."

At the moment there is no definitive answer to the question whether any one of the ancient gnomonists ever built a dial with its style directed towards the pole. It is clear that he would have had no following because the hours of his dial would not have agreed with those of the dials in use, whose hours were 'temporary.' But it is worth mentioning here that there is in the Naples Museum a mosaic originating from Pompeii, therefore prior to AD 79, in which a dial with an inclined style has been tentatively recognized in the background. The rare passages in the written records on the nature of the known dials are not very ex-

plicit when it comes to details. Pliny reports that the scaphion was improved by Anaximander and by Eratosthenes (276–195 BC). Eratosthenes would have used a scaphion in order to determine the difference in latitude between Alexandria and Syene, an operation which leaves no doubt of his knowledge of the earth's sphericity. Eudoxius of Cnidus devised a dial provided with curves to which the name of arachnid had been given. It was inferred therefrom that it possibly bore drawings of declination lines, as these lines resemble a giant spider when laid on a flat dial. Eudoxius would, then, have been the first to draw these curves on a dial. It was not until the time of Julius Caesar that there appeared more than an incidental mention of the sundials of the era.

The book, *De architectura*, by Vitruvius, an architect and war engineer to whom we also owe the description of the clepsydra, contains a list of the then known dials with the name of the presumed inventor affixed to each. There are thirteen such dials: the hemicyclium of Berossos, the scaphion of Aristarchus of Samos, the disc *in plano* of Aristarchus, the arachnid of Eudoxius of Cnidus, the plinth of Scopus of Syracuse, the *prosta istoroumena* of Parmenion, the *pro pauditnia* of Theodosus, the *pelecinon* of Patrocles, the cone of Dionysidorus, the quiver of Apollonius.

Vitruvius concludes his list by stating that "these men and others have left us all kinds of dials, such as the conarachnid, the eugenaton and the antiboreum." He attributes the invention of the scaphion to Aristarchus of Samos, and Aristarchus did indeed devise dials of this shape; but since Berossos lived a century earlier and since the scaphion is only an external modification of his dial, it is clear that credit for the scaphion rightly belongs to Berossos.

Although we do not know most of the dials mentioned by Vitruvius, we may infer some hypothetical features for some of them from their names. The probable appearance of the arachnid has already been discussed. It is thought that the disc could be this type of horizontal dial, a sample of which has been found along the Appian Way. The vertical gnomon of this dial (see figure 9) is located at M. When the AB axis is properly oriented, with point M in the south, the shadow of the tip of the gnomon follows the two hyperbolas of the solstices, which have been traced for the latitude of Rome. Hour lines have also been traced in such a way that equal intervals of time are delineated. Empirical observations were doubtlessly required to determine for each line two points which were later joined. Aristarchus may possibly have possessed the astronomical and technical knowledge required for the construction of this dial. Figure 9 shows the dial surrounded by two circles; the dial found on the Appian Way carries within these circles the names of the winds Desolinus, Eurus, Auster, Africus Faonius, Aquilo, Septembrio, Boreas, as well as the signature "Antistius-Euporus fecit."

As for the cone, we have seen that it developed from the hemicyclium (figure 10) without any essential differences. The tip of

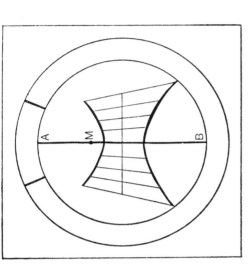

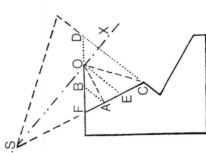

FIGURE 9   What we assume to be the *discus in planitia* of Aristarchus of Samos was found in the Vigna Cassini near the Appian Way at the beginning of the nineteenth century. The figure is taken from the book *Di un antico orologio solare* by Francesco Peter (Rome, 1815)

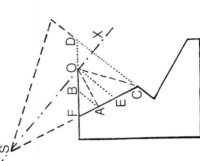

FIGURE 10   The so-called Cone of Dionysidorus derived from the hemicyclum. In the section, the triangle DFC stands for the cut of cone in the stone. The tip of the cone is S, FO is the style whose end O lies on the axis SX of the cone. Each of the angles AOE and EOC is equal to the largest declination of the sun. AB, EO, and CD are the projections of the circles of the solstices and of the equator

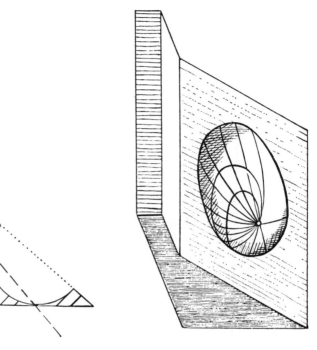

Section

FIGURE 11 The inventor of the antiboreum has found a way of making the indications of the sun more luminous by using a bright spot in a shadowy hemisphere. However, this leads to the complication of finding the intersection of conical and spherical surfaces in order to trace the lines of the solstices. The path of the light ray is shown in the section

the gnomon was located in the axis of the cone in such a way that the daily circles drawn by its shadow were located in planes perpendicular to the axis. Figure 10 shows the construction of the circles for the solstices and the equinox.

Finally the enigmatic antiboreum has been linked to a dial of the scaphion family and of Roman origin in the Berlin Museum. It is a hollow hemisphere cut in a block of stone (figure 11) pierced by a little hole in the bottom. In contrast to the other scaphions the opening of the hemisphere was oriented towards the North. A thin bundle of the sun's rays entered the sphere through a little hole leaving on its walls a small image of the sun, which indicated the hours. Since in the Berlin scaphion the hour-lines do not indicate the old temporary hours but equal astronomical hours, this dial possibly belongs to a more recent era.

It is interesting to note in Vitruvius' book a remark that leads to the conclusion that there were many books in existence dealing with the construction of portable dials and that this construction was accessible to all who knew the analemma. (This Greek word is still in use and the reader will find it in chapter six, where the very special type of dials which carry its name will be studied.) The analemma was used in the days of Claudius Ptolemy, in the second century AD, to designate a type of simplified armillary sphere* which later became the astrolabe. It was applied to a very original technique used by the Ancients to resolve, with the help of a small *épure*, astronomical calculations in the field of spherical trigonometry. Today it is interpreted as an orthographic projection and is used in this sense in modern gnomonics. The meaning assigned to it by Vitruvius is not clear, but it seems that one needed only a minimal amount of astronomical knowledge to "know the analemma." We must conclude by necessity that in

* A construction of rings as used by ancient astronomers to show the celestial equator, tropics, etc.; a skeletal celestial sphere.

Vitruvius' day there were no practical methods available for the construction of ordinary dials. The renaissance of the sundial towards the end of the Middle Ages initiated these practical methods.

Elsewhere in his book on architecture, Vitruvius speaks of the tower of Andronicus in Athens, now called the Tower of the Winds (Plates 2 and 3). An octagonal tower, dating from the first century BC, it stands in its entirety today, the top of each side being decorated by a relief representing one type of wind. Vitruvius does not mention the presence of sundials on the walls of the tower, yet on each face, immediately under the reliefs, very old lines drawn in the stone represent sundials still bearing their styles. The styles are iron rods placed perpendicularly in the table of the dials and carrying at their tips small spheres, whose shadows give the reading of the dial. Nothing is known about the origin and the age of these dials, which have been studied by Delambre, but it is almost certain that they are considerably younger than the tower itself.

The *Opus agriculturae* by the Latin agronomist Palladius, who lived during the fourth century and owned vast agricultural properties in Sicily and in the district of Naples, contains a section dealing with sundials or, to be precise, with the gnomon. It is somewhat of a surprise to discover on reading the sections on gnomonics in this book, which was meant primarily for farmers, that some centuries after Vitruvius' list of dials made about the time of Plautus when the cities were filled with them, the countryside for the most part still knew only the gnomon. The fact that this primitive instrument could be used without special instrumentation and in any location may explain this lack of knowledge. Palladius quotes in his work one of those tables, then widely used, which gives the length of the shadow, varying with the seasons, for the hours of the day. Such tables were approximate; they were calculated for a given latitude and a new one was needed for each

month. For example, the Palladian table, for January, valid in Sicily, looked like this:

Hora I et XI	Pedes xxix
Hora II et X	Pedes xix
Hora III et IX	Pedes xv
Hora IV et VIII	Pedes xii
Hora V et VII	Pedes x
Hora VI	Pedes ix

The shortest shadow, nine feet, corresponded to the hour of noon, i.e., 6 o'clock or the sixth hour in "temporary" hours. Each length gave two hour values, one for the morning and one for the evening respectively, at equal intervals from noon. The tables could be accurate only for a few days of the month because of the continuous variation of the solar declination. But they could not have been meant to be accurate, since Palladius rounded off the distance figures to whole numbers which, of course, were easier to remember.

Such tables, originating in the fifth and sixth centuries, have been found on the walls of an ancient temple in Tehsa, in Nubia, not far from the old Syene, the Aswan of today. The English monk, Bede, had set up similar tables in his book *De mensura horologii* early in the eighth century.

As the Roman empire declined, so too, it seems, did the use of sundials. This interpretation is supported by the very sporadic character of the gnomonic finds from the first centuries of the second millennium of our era.

By around the year 1000, the Arabs had become the inheritors of Greek gnomonics, as well as of all the other ancient sciences. Fifteen of their books on gnomonics written during the period from the eleventh to the fourteenth century have survived. The title of one of them, the work of Ali Abul Hassan, a Moroccan scientist of the thirteenth century, seems to deal with a dial whose style was oriented towards the pole. The same author speaks among other things about hour-lines for equal hours, a division of time similar to ours. He used solar elevation and that of the circumpolar stars during the days of the solstices to evaluate the elevation of the pole; he was perhaps the first to use eclipses of the moon to calculate longitude.

After the Crusades, sundials, with their styles directed towards the pole, i.e., parallel to the axis of the world, suddenly appeared over most of Europe. It is wholly probable that the contacts of the Crusaders with the Arabs accounted for this unheralded appearance, but the name of the inventor of this new type of dial is unknown. This type is the classic dial now everywhere in use (except for the analemmatic dial, also a precise and mathematically exact instrument, which is mostly found in France). The scarcity of the old type of dials in Europe during the days of the Crusades could be explained by the fact that the novelty of the classic dial partly involved their replacement and even systematic destruction. Thus there began an entirely new era in gnomonics. There was nothing left to invent: the sundial had reached perfection and taken its final form, though its practical realization still left the field wide open to the imagination. From the Renaissance until today, the makers of sundials have shown an astonishing originality. Garden sundials have been traced on all types of surfaces and the portable dials have appeared in widely varying forms – as flat or round boxes, spheres, cubes, and even as jewels to be set into rings.

The era of the Renaissance was also that of the discovery of America. But only later was it learned in Europe that, like the ancient people of the Orient, the Incas and the Aztecs had been using the gnomon for centuries for the observation of the solstices and the equinox, and that their knowledge of the calendar was unsurpassed.

The birth of the mechanical watch with wheels should have sounded the immediate death knell of the sundial. Not so, for the watch had chronic imperfections. Its use accelerated progress in the division of time in daily life, but its need for frequent setting made the sundial as essential as ever and has even contributed to its modern form, the heliochronometer. This is a kind of equatorial dial consisting of a wire stretched parallel to the axis of the world (figure 12). It is not generally known that this instrument was used into the twentieth century by some networks of the French railways for uniformity in the setting of the station clocks.

This need for setting, more recently confirmed by the telegraph and wireless networks, gave rise in earlier days to other installations, often set up in the cathedrals, called noon-mark dials. These supplied the exact time at the moment of the solar passage at the meridian, i.e., the true noon. The light of the sun came through a hole pierced in the southern façade of these buildings as a thin shaft. The trace of the plane of the meridian was then drawn on the floor; the image of the sun, by crossing this meridian line, indicated the hour of noon with an accuracy that depended on the dimensions of the structure but which was within the second. These noon-mark dials also supplied the dates of some non-movable ecclesiastical holidays. The best known are those of St Sulpice in Paris, that of the cathedral of Florence, which will be mentioned later, and that of St Petronius in Bologna.

Today, as we well know, the sundial has lost, one by one, the practical uses which served our ancestors for so long. Nevertheless, it remains a living, beautiful, and instructive reminder of the past, a witness to the centuries and to life now gone. It invites rest and contemplation; through a hidden and primeval call to our subconscious, it evokes mysteries and echoes of the distant and forgotten times when man took his first steps towards a more cultured life.

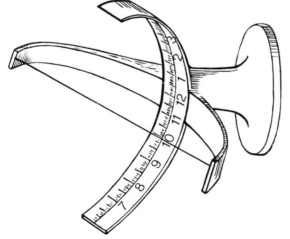

FIGURE 12 Few people know that the heliochronomometer was used until 1900 to regulate the time of some railroad stations of the French network. The reading required the correction for the longitude and the equation of time. The time could be read to the nearest minute

CHAPTER TWO

Some Gnomonic Cosmography

1 USEFULNESS

It is not necessary to be a great mathematician or a great astronomer to understand the principles behind the working of a sundial. As we shall see, it is possible to construct one without any theoretical knowledge, by using some of the rules given in chapter three.

A minimum knowledge of the relative motions of the sun, earth, and moon will prove useful, however; its acquisition, even from this chapter, prudently limited as it is, can be profitable for future friends of gnomonics. A few hours will suffice to gain an understanding of the main celestial motions. So interesting are these phenomena that the informed person cannot fail to be astonished by the lack of interest in them shown by large numbers of people. We beg the indulgence of the informed, but we feel that a short summary of the facts is useful, if not necessary.

All forms of life on earth depend entirely on the presence of the sun and are regulated by it. The sun rises and sets with infallible regularity, the rhythm of the seasons is immutable, and the moon, our neighbour – about which one speaks more often now – follows quietly its daily or nightly course, in sight of all on the earth. The immense spectacle of the heavens has imposed itself on the atten-

tion of man since the beginnings of history, and astronomy, the most ancient of man's sciences, has been rightly called the mother of all the others.

The secrets of the sundial will vanish for the friend of gnomonics as he becomes more familiar with those of the heavens. When he has assimilated its essential principles, he will look at the sundial with infinitely more satisfaction and profit: through the shadow of the style between the hour-lines, he will visualize the oblique course of the sun along the ecliptic; the style, a simple rod, will look to him like a magic line around which the celestial bodies of the whole universe rotate in full mathematical rigour, in planes perpendicular to its axis.

2 GEOGRAPHICAL CO-ORDINATES

The earth is a body whose shape is almost that of a sphere slightly flattened at the poles. The flattening would allow us to consider it as an ellipsoid of revolution; i.e., a body whose section by a plane containing the axis of the poles would be an ellipse, if the internal

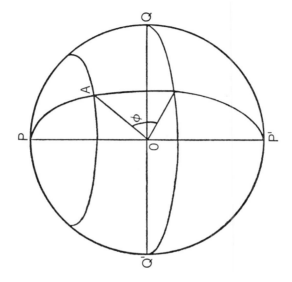

FIGURE 13

masses which make it up had the same density everywhere and if they were submitted to a uniform pressure. Since this is not so, the real shape of the earth turns out to be more complicated; it cannot in fact be defined analytically, so that the elaborate calculations which have been carried out to express it have led to complicated formulae. Finally the word *geoid* was coined to designate the shape.

In explaining the elementary principles contained in this chapter, we shall, for practical reasons, neglect the flattening of the terrestrial globe and consider it as a perfect sphere: this will lead to no inconvenience.

Everyone knows that the earth rotates and that it completes a revolution around the axis of its poles in twenty-four hours. During the course of this rotation any point A of its surface will draw a circle whose plane is perpendicular to the axis. The circle is called the parallel of A and it is determined by the angle φ subtended by the radius OA and the plane of the equator (figure 13). φ is then the *latitude* of the point A and of all the other points located on its parallel. From the geographical maps we know that the latitude is measured in degrees from zero to 90, starting from the equator, that it is north (or positive) for points in the northern hemisphere, and south (or negative) for those in the southern hemisphere. The latitude is zero at the equator. The equator is also a parallel; it has, however, the peculiarity of being the only one with the same radius as that of the earth and to be therefore what is called a great circle.

The *vertical* at A is the perpendicular to the surface of the globe at this point, assuming the surface to be as perfectly flat, regular, and horizontal as that of a liquid would be. Such a plane is called the horizontal plane or simply the *horizon* when it is assumed to extend indefinitely in all directions. The plane of the horizon – a term that we will use frequently – is obviously tangent to the surface of the globe. The fact that it is not quite so in reality, because of the flattening of the poles, will not invalidate our considerations. The vertical at one place, therefore, makes an angle with the plane of the equator which is equal to the latitude.

On the other hand, half a great circle with the line of the poles PP' as the diameter is called a *meridian*. Any point A on the surface of the globe will be on a meridian. The particular meridian that passes through the observatory of Greenwich, England, has been chosen as the origin or first meridian, the o (zero) meridian. For a given point A, the longitude is defined as the difference, measured in degrees, between the meridian of A and the origin. The longitude will be east or west and its values will lie between o (zero) and 180°. The location of any point on the globe will, therefore, be determined by its latitude and its longitude, i.e., by its geographical co-ordinates.

Let us add that the longitude may be expressed in units of time because a 360° rotation of the globe corresponds to a duration of twenty-four hours. Table 1 shows the conversion of longitude from degrees into hours or *vice versa* using this basis of equivalence.

3 THE EARTH AND ITS MOTION

In the previous paragraph we said that the earth rotates around its axis. This rotation, combined with the illumination of the sun results in the alternation of day and night. On the other hand, the earth takes one year to complete one revolution around the sun. One focus of the elliptic orbit followed by the earth is occupied by the sun. The time interval separating two of its consecutive passages across a given point of the orbit has been calculated as one *tropical year*, with a duration of 365.2422 days.

The axis of the earth's rotation is not perpendicular, but is inclined at an angle of 23° 27', to the plane of its orbit. In addition, rather than staying rigorously parallel to itself, the axis of the earth undergoes a small retrograde motion around the vertical to the orbit; this motion is so slow that it takes a full 26,000 years to complete a cycle. Nevertheless, it is fast enough for the poles of the heavens located on the extensions of the earth's axis to have been displaced appreciably since the beginning of history. It indeed represents an annual motion of about 50 seconds and is the basic explanation of what is called the *precession of the equinox*. However, we shall disregard it in our explanation of figure 14 where s stands for the centre of the sun, while A, B, C, and D are four typical positions of the earth on its orbit.

A and C have been chosen in such a way that s lies in the plane of the terrestrial equator; likewise, if T is the centre of the earth, the radius TS of the orbit is perpendicular to the earth's axis in A as well as in C. In B and D, however, it makes the maximum angle of 23° 27' with the plane of the equator in the direction indicated by the figure.

We should note here that the distance of the sun is so enormous that its rays may be considered parallel in an area of space with the dimensions of the earth. Under this assumption, since the rays will be in the plane of the equator, they will be tangent to both poles at A and C.

There will be an equality then at A between day and night over the whole surface of the globe, and on the equator the solar transit in the meridian will take place at the zenith. This situation occurs every year on March 21, when the spring *equinox* occurs.

When the earth moves farther away from A towards B, the radius ST leaves the plane of the equator and is inclined to it at an increasing angle which reaches its maximum value of 23° 27' at B (figure 15). The rays of the sun are tangent to the terrestial

PLATE 1 Roman dial of the second century AD found at Bettwiller (Alsace) in 1879. This valuable carving, 53 cm high, made of local sandstone, constitutes one of the jewels of the Archaeological Museum of Strasbourg.
(Photograph: Muckensturm, Strasbourg)

PLATE 2 The Tower of Andronicus in Athens. It is now called the Tower of the Winds and it is one of the best preserved monuments of Greek antiquity. The straight styles of the dials may be distinguished under the figures on the reliefs.
(Photograph: V. and N. Tombazi, Athens)

PLATE 3   Details of the dials on the Tower of the Winds in Athens. The indications were given for the shadow of the tip of the straight styles. Nothing is known about the origin and the age of these dials. (*Photograph: V. and N. Tombazi, Athens*)

PLATE 4   The Adolescent with the Sundial of the Cathedral of Strasbourg. It has faced the Place du Château for five centuries. Its mediaeval dial with temporary hours is one of the most primitive left to us from that era. (*Photograph: Strasbourg Museum*)

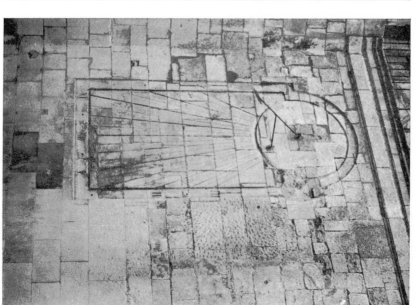

PLATE 5   Dial in the court of the City Hall in La Rochelle, fifteenth or sixteenth century (*Photograph: the author*)

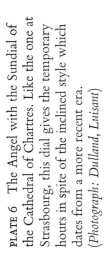

PLATE 6 The Angel with the Sundial of the Cathedral of Chartres. Like the one at Strasbourg, this dial gives the temporary hours in spite of the inclined style which dates from a more recent era. *(Photograph: Dulland, Luisant)*

PLATE 7 The Man with the Sundial of the Cathedral of Strasbourg. It dates from 1493 and its dial is the earliest known for which the hours are no longer temporary. *(Photograph: Strasbourg Museum)*

8

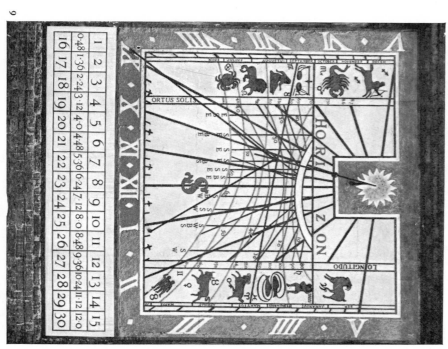

9

PLATE 8 Small stained-glass window dating from 1731, kept in the Historical Museum of Basel and carrying a dial designed for a style installed on the exterior. In the interior, the reading of the hours was done clockwise.
(Photograph: Basel Historical Museum)

PLATE 9 The celebrated dial of Queens' College. It looks very complicated at first sight but its reading is rather easy owing to a judicious choice of colours for its numerous lines.
(Photograph: Edward Leigh, Cambridge)

PLATE 10 Modern vertical dial in pink moulded concrete. The dial carries the Babylonic and Italic lines.
(*Photograph: the author*)

PLATE 11 Another modern dial in moulded concrete. The dial, which is almost occidental, carries the same information as the one in Plate 10. However, a graph has been added to make it usable as a moon dial.
(*Photograph: Moreau, Carcassonne*)

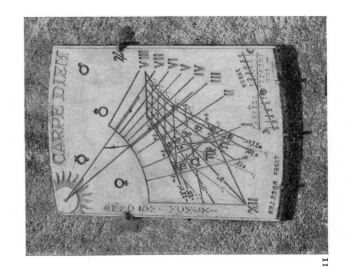

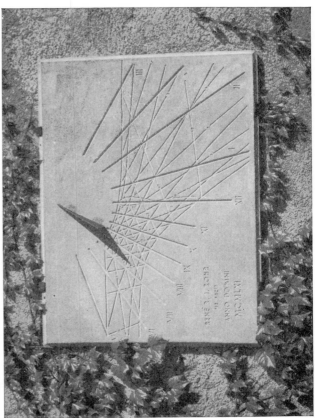

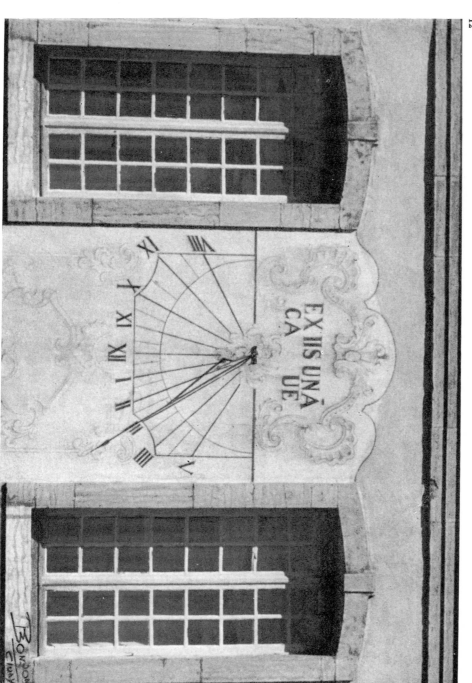

PLATE 12 The concise, simple, and
almost meridional dial of the ancient
Benedictines of the Abbey of Cluny carries
the quotation: "Beware of one of them."
(*Photograph: Pierre Bonzon, Cluny*)

12

globe along a great circle which delineates day from night. In following the daily path of a point M located in the northern hemisphere, it will be noticed that the portion of the parallel followed in the region illuminated by the sun is longer than that on the dark side. In other words, the days will be longer than the nights in the northern hemisphere, while the reverse will hold in the southern hemisphere. The situation will be changed slightly after the departure of the earth from A, but there will be a progressive prolongation in the days north of the equator and a shortening south of it. The trend will be reversed at B and we shall end up in C with a situation analogous to that found in A.

Position B is that of June 21, i.e., the summer solstice. While

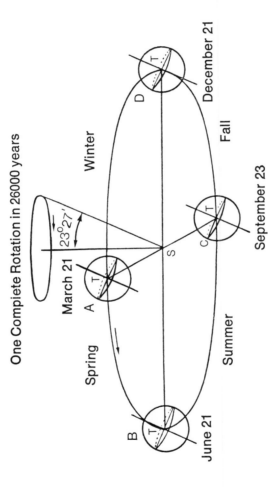

One Complete Rotation in 26000 years

Spring

March 21

A

B

June 21

Summer

S

23°27'

Winter

D

December 21

C

September 23

Fall

FIGURE 14

spring will have lasted all through the earth's trip from A to B, summer will last from B to C. June 21 will be the longest day of the year in the northern hemisphere and the shortest in the southern hemisphere, since the angle between the radius sт and the plane of the equator in B reaches its maximum value there. At c, September 23, it is again the equinox, but the days go on

varying in the direction already initiated. By studying figure 14, the reader will be able to follow the complete cycle of the seasons till the reappearance of point A on March 21.

A combined study of figures 14 and 15 shows that the equinox lasts all year round for inhabitants of the equator while, at the poles, day and night last six months each. We see that the inclina-

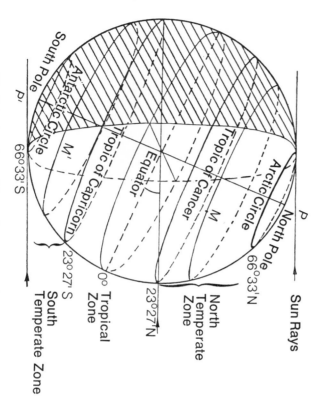

FIGURE 15   The lighted regions of the earth on June 21 (summer solstice)

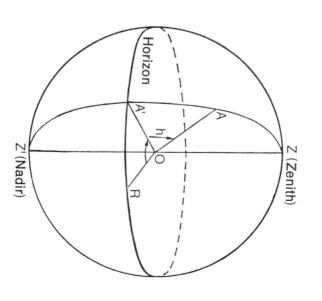

FIGURE 16

tion of the earth's axis on its orbit has a major influence on life on earth since it causes the succession of the seasons and the alternation of climates and therefore determines the suitability of the earth for habitation as far as the high latitudes.

In addition, the phenomena just described cause some peculiarities in certain parallels. Indeed, looking at figure 15 (which corresponds to the position B of figure 14) we notice that on June 21 the sun does not set during twenty-four hours over all the points of the earth located north of the 66° 33' N parallel. This parallel, 23° 27' distant from the pole, is called the Arctic Circle. The meridian transit will occur at the zenith all along another parallel, 23° 27' N, called the Tropic of Cancer. On December 21 (position D in figure 14), the same phenomena occur at the Antarctic Circle (66° 33' s) and the Tropic of Capricorn (23° 27' s). These parallels delineate on the globe the tropical, temperate, and polar zones.

Finally let us mention here that the point of the earth's orbit corresponding to A in figure 14 is called the *vernal equinox*. It is usually designated by the letter γ. We shall refer to it again when the equatorial co-ordinates are discussed. Returning to the definition of the tropical year, we can now define it precisely as the time interval between two consecutive passages of the sun at the point γ. This point is not fixed, but retrogresses on the orbit by 50" of arc per year, so that it takes 26,000 years to complete the cycle back to the point of departure. The line between the two equinoxes rotates with it. This is the so-called precession of the equinoxes.

4 LOCAL CO-ORDINATES

The representation of the firmament by a drawing is the first of the difficult tasks to face the uninitiated who wish to study problems of astronomy.

The surface of the earth looks to the beginner like a horizontal plane limited by the horizon, while around and above him the sky looks like an immense hemispherical vault. He cannot help but imagine that the hemisphere becomes a sphere below the horizon. He can then represent it by a circle, of which he is the centre. The plane of the horizon can be represented by another circle seen in perspective, i.e., an ellipse. This drawing and what it represents is commonly called the *local sphere* of the observer. The point located on the vertical above the observer is the *zenith*, the opposite point on the sphere is the *nadir* (figure 16).

In order to represent the image of a star on the local sphere, it is necessary to picture mentally a system of co-ordinates made up of the elevation of the star above the horizon and of its orientation, measured from a fixed direction, be it north, south, or any other.

These two measurements are called the *horizontal co-ordinates* or sometimes the *horizontal co-ordinates* of the star. Taking A as a star on the local sphere, the plane zAz' is called the vertical plane of the star and the angle *h* between the radii OA' and OA is its elevation measured in degrees. It is useful at times to know the angle zOA, called the *zenith distance* of the star, which evidently equals (90° − *h*). To complete the determination of the position of A, it suffices to measure the angle ROA' of the direction of the vertical plane of the star with the reference direction. This angle is called the *azimuth* and is measured in degrees, counterclockwise. In practice, the south is taken as the origin in azimuthal measurements. A description of how one proceeds to measure this direction follows:

When any star is observed and its hourly co-ordinates are noted as it appears to move across the sky it will be seen that it draws an arc of a circle on the local sphere, between rising and setting, analogous to that of the sun (the star should preferably be chosen in a direction south of the east–west line which is always known approximately). The star rises during the first half of its trajectory and descends during the second half. During each of these halves, it will pass through a certain elevation h. If the azimuth is measured at the precise instant of its passages at this elevation, the direction of the true south will be obtained by taking the mean of the two measurements. If, for instance (figure 17), OR is the reference direction and A_1 and A_2 are the azimuths of the star at the two instants when its height is the same in the first and second halves of its trajectory, the azimuth or the direction of the true south, i.e., of the direction of the meridian measured with respect to the direction OR, will be $Am = \frac{1}{2}(A_1 + A_2)$. The mean of the previous observation times yields the time of the transit of the star through the meridian.

If we compare the times of the meridian transits of a star that is not the sun or the moon or a planet, we realize that the interval of time between two consecutive transits will be constant and that it will equal approximately 23 hours and 56 minutes. This time lapse is called *sidereal day*. One could do the same observations with other stars in the direction of the north.

There, however, it will be found that some stars never set: they seem to describe circles around a point very close to the Polar Star. The point is the North Pole of the celestial sphere, whose counterpart in the other half of the sky is the South Pole. Such stars are called *circumpolar stars*.

If the elevation of a circumpolar star is measured on both sides of the meridian, the mean of the measurements is the elevation of the celestial pole. This angle, as we shall see, equals the latitude of the point of observation. The straight line joining the two celestial poles is called the *axis of the world* because it appears to serve as an axis of rotation for the whole celestial sphere.

The plane perpendicular in O to this axis is called the *equator* (figure 18). The celestial sphere will be separated in this way into two hemispheres with poles and equator, exactly like the earth or a geographical globe. Thus, we know how to determine on our local sphere the direction of the south and of the north as well as the east–west axis which is perpendicular to it.

If Q'Q is the plane of the equator and z the zenith, the angle

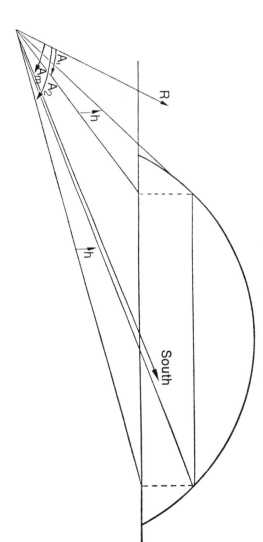

FIGURE 17

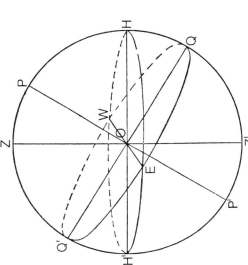

FIGURE 18

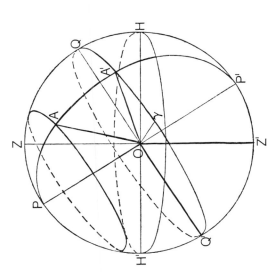

FIGURE 19

Q'OZ will be defined as the latitude. The angle is equal to POH or, as we have just stated, the latitude of a point is equal to the elevation of the pole above the horizon at that point.

Figure 18 is the most convenient representation for the study of the celestial sphere. Naturally, the universe actually does not rotate around the axis of the poles. The rotation is an illusion created by the rotation of the earth. However, in order to keep the explanations simple, we shall consider the earth as fixed and the universe as endowed with the movement of rotation. This assumption alters none of the relationships though we shall also keep in mind what is really happening, and the result will be the same.

5 HOUR AND EQUATORIAL CO-ORDINATES

The horizontal co-ordinates defined above have the very serious inconvenience of referring to the zenith and to the horizon which themselves are not fixed reference points on the celestial sphere. Figure 19 suggests to us the more logical solution of choosing the pole and the equator, which are fixed. In figure 19 the hour-circle of a star A is the half-circle PAP' and the hour-angle of the star is the angle formed by its half-circle with the half-circle PZP', which is the local meridian. The hour-angle, measured on the equator, will then be the angle QOA', A' being the point of intersection of the hour-circle PAP' with the equator. The hour-angle is measured in degrees in a retrograde direction (the direction of the daily motion), that is, it increases monotonically, starting from the stellar transit on the meridian. We have already mentioned that the hour-angle takes approximately 23 hours and 56 minutes, i.e., one sidereal day, to increase from zero to 360°, or, in other words, to complete a revolution. The sidereal day is divided into twenty-four sidereal hours. Therefore, the hour-angle increases by

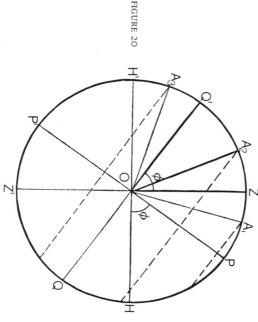

FIGURE 20

$360°/24 = 15°$ in one hour. Thus the hour-angles may be measured in hours and minutes; and this, indeed, is the customary way in which it is done.

The *declination* of a star is its angular distance A'OA from the equator, which is positive or negative depending on whether the star lies to the north or the south of the equator. Although the declination is a co-ordinate independent of the daily motion, the hour-angle, which is measured from the meridian, is not; we have seen that the hour-angles increase as time goes on.

The choosing of a fixed point on the equator, the vernal equinox, usually designated by γ, has already been considered. Then what is called the *right ascension* of the star A is the angle γOA' measured in hours and minutes in the positive direction.

The advantage of this definition is obvious. Indeed, so-called astronomical clocks are to be found which indicate a time lapse of 24 sidereal hours when 23 hours and about 56 minutes of our ordinary time have lapsed. Such a clock, therefore, runs through 24 hours between the two meridian transits of a fixed star. (We recall that, according to our definitions, the sun is not a fixed star.) In addition, these astronomical clocks have been set in such a way that they indicate zero hour at the instant of the meridian transit of the vernal point. From then on, the right ascension of a star is automatically indicated by the astronomical clock at the instant of its meridian transit.

In principle – although some restrictions must be imposed – an ordinary clock performs in an analogous way for the sun, thus indicating its hour-angle.

The measurement of the right ascension of a star is extremely simple. It is not so easy for the declination. Here, the elevation of the star is first to be noted at its meridian transit. In a location of latitude ϕ, either of the two following cases will apply (figure 20).

(1) The star reaches its highest point at the meridian between z

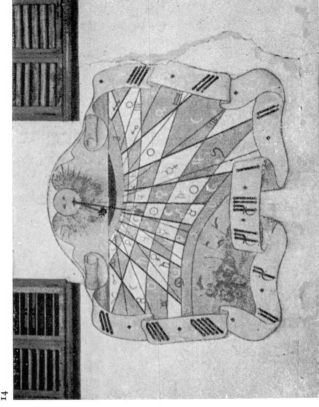

PLATE 13 The dial of the Church of Telfes near Innsbruck (Austria), one of the best preserved dials of Peter Anich. It dates from around 1760. The vertical bands give the length of the day and the elevation of the sun at noon.

(*Photograph: Dr Kühnelt, Innsbruck*)

PLATE 14 This dial with planetary hours at Prutz in the Tyrol is one of the most peculiar dials in Europe with its table of the Masters of the Day (figure 91) and its accessory division in temporary hours.

(*Photograph: Dr Kühnelt, Innsbruck*)

PLATE 15 Another Tyrolean dial worthy of interest is that of the Castle of Ambras near Innsbruck. Its presentation under the necklace of the Golden Fleece surrounding the arms of the lords of the castle and its WNW orientation make it a gnomonic curiosity. We note indeed that the hour-lines – almost parallel – converge towards the bottom right of the dial. (*Photograph: Dr Kühnelt, Innsbruck*)

PLATE 16 Dial by Peter Anich on the
Church of Natters in the Tyrol. Vividly
coloured and highly decorative, it dates
from 1759. The Roman numerals are
reproduced partially in Arabic numerals
along a band following the path of the arc
of the summer solstice.
(*Photograph: Dr Kühnelt, Innsbruck*)

PLATE 17 Seventeenth-century engraving
showing the layout of the meridian
installed by Toscanelli under the cupola of
the Cathedral of Florence

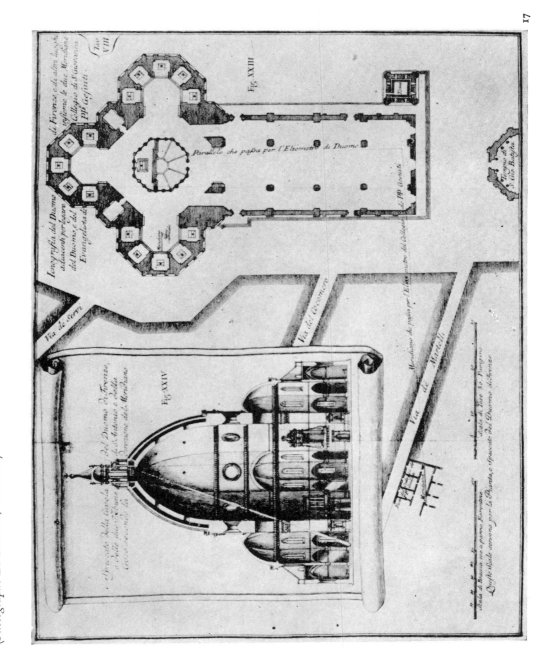

PLATE 18 The triple dial of 1572 decorating the front of the southern transept of the Cathedral of Strasbourg. To the left, the elevation and azimuth of the sun; to the right, the Babylonic and Italic hours. Old Gothic numerals. The dial on the right carries the quotation: *Tempus edax rerum,* which is not reproduced on the engraving. On the right, we read: *Veritas temporis filia. (Photograph: Strasbourg Museum)*

and P, i.e., to the north of the zenith. Its declination is $\angle Q'OA_1$. But $\angle Q'OA_1 = \phi + (90° - h)$, namely, the sum of the latitude and the zenith distance.

(2) The culmination takes place to the south of the zenith. $\angle Q'OA_2 = \phi - (90° - h)$, namely, the difference between the latitude and the zenith difference. We note that in the case of star A_3 the difference will be negative. The declination will then be negative. Indeed, A_3 is located below the equator.

Let us recall here the very important fact that the refraction of light rays by the earth's atmosphere causes the stars to appear closer to the zenith. The observed elevations must therefore be corrected for the refraction effect which is itself dependent on the air temperature and on the barometric pressure. Table 2 lists the mean values of refraction for various observed elevations.

6 THE SUN

It is universally known that the sun rises in the east and sets in the west after having reached its highest elevation at true noon at the instant of its passage through the plane of the meridian. During this time interval, as mentioned earlier, the shadow on the ground of the objects illuminated is very long as the sun rises, shortens during the morning, goes through a minimum at true noon, and lengthens again as the sun sets.

We have also stated that, for an observer on the earth, everything seems to happen as if the earth were immobile and the sun rotated around it. This was the official belief in Europe until the days of Galileo.

In contrast with the sidereal day, which is the time interval between two consecutive meridian transits of a star, a *true day* is meant to represent the time interval between two solar transits through the meridian. The true or solar day is longer than the sidereal day by four minutes.

If the position of the sun, measured in equatorial co-ordinates, is plotted daily on a map of the sky, we realize that it draws a complete circle on the celestial sphere during the course of one year. In the northern hemisphere we must face south to observe this phenomenon; the daily displacement will therefore be towards the left. Every day, at its meridian transit, the sun will be a little bit late compared with the stars in its vicinity. This is the reason for the difference between the true day and the sidereal day. If the sun happens to reach its highest elevation at the same time as another star, the meridian the next day will have to have rotated for about four minutes more in order to catch up with the sun. These four minutes represent an average value since the speed of displacement of the sun is not constant.

The Ancients were long aware of this fact, and the Chaldeans noted the constellations crossed successively by the sun, limiting their number to twelve: Aquarius, Pisces, Aries, Taurus, Gemini, Cancer, Leo, Virgo, Libra, Scorpio, Sagittarius, Capricorn, to name them in their natural order.

We must admit that the division has no practical use. Some purpose is served by knowing it, however, since the symbols for the twelve signs are valued in gnomonics for the decorations of the dials or to indicate pictorially the positions of the sun in the sky; indirectly it may also indicate (we shall speak of this again) the declination of the sun and therefore the season of the year.

The sequence of the twelve constellations is called the zodiac, from the Greek word *zodios* meaning "figure of an animal", in spite of the presence of some anthropomorphic signs in it and even that of a scale. The Greek scientist Hipparchus put some order into the zodiac by dividing it into twelve equal sections of 30° each, but, for the sake of convenience, he kept the old names. We have mentioned previously the precession of the equinoxes, a phenomenon first noted by Hipparchus and the effect of which, since his lifetime, has been gradually to shift the constellations in the sky away for the signs which now carry their names. Figure 21 shows the apparent trajectory of the sun on the celestial sphere with the zones of the constellations as they appeared in the days of Hipparchus.

The study of the positions of the sun in the sky, or more simply, of an almanac giving the right ascension and the declination of the sun for each day of the year leads to the following conclusions:

(1) The right ascension increases daily by approximately one degree (four minutes of time). The increase is the result of the daily displacement of the sun on the celestial sphere.

(2) The declination is zero on March 21, after which time it increases. It goes through a maximum value of 23° 27′ on June 21, then decreases, and reaches zero again on September 23; it goes on decreasing down to the minimum value of −23° 27′ on December 21 and then starts increasing again until it reaches zero on March 21. This cycle is repeated year after year.

The plot of the co-ordinates of the sun during the course of a whole year makes it clear that a great circle is described called the *ecliptic*, the plane of which is inclined by 23° 27′ with respect to the equator. We have already encountered this same fact in another guise (see figure 14).

The ecliptic is the central circle of the zone occupied by the signs of the zodiac. The point γ mentioned in the previous chapters is the intersection of the ascending branch of the ecliptic with the equator. It is worth recalling that it retrogresses by 50″ per year on the equator because of the precession of the equinoxes. The time interval that elapses between two solar passages through this point is our calendar year – the tropical year which, we repeat, has a duration of 365.2422 days or 365 days 5 hours 48 minutes and 46 seconds.

The sun goes through the point γ at the moment of the spring equinox. Each day thereafter it will fall progressively about four minutes behind this point; when it has caught up with it once again, one year will have elapsed. During this time, the vernal point will then have gone through the meridian once more than the sun and the sidereal year will, therefore, be longer than the tropical year by one day. The duration of the sidereal day, measured in mean hours, will then be

$$365.2422/366.2422 = 23 \text{ hours } 56 \text{ minutes } 4 \text{ seconds.}$$

We find here once again the difference of approximately four minutes.

7 THE MEASURE OF TIME

The sidereal day is the time taken for a complete rotation of the celestial sphere. It is defined ultimately as the time interval separ-

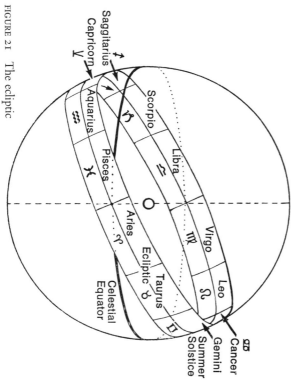

FIGURE 21   The ecliptic

ating two consecutive passages of the point γ through the local meridian.

Sidereal time, so-called, is the time elapsed since the meridian passage of the point γ. This implies that the sidereal time equals the hour-angle of γ, and, as already noted, it differs from our clock time. But we have also seen that astronomical clocks have been specially built in observatories to follow it.

It would, however, be inconvenient to regulate our life on sidereal time. Since it is the sun that regulates our activities, a *true solar time* or *true time* has been defined as the hour-angle of the sun starting from noon. This is the time indicated by our sundials, using, so to speak, the sun directly. It has the disadvantage of not being rigorously constant. Indeed, the apparent orbit of the sun is an ellipse and the plane of the orbit is inclined with respect to the equator; the hours of true time are not equal in length throughout the year. In order to have hours of constant duration and to make possible the rational use of mechanical clocks, a *mean solar time* or *mean time* has been devised. This mean time is the time of a fictitious sun or *mean sun* progressing at a constant speed around the equator instead of around the ecliptic; it would leave the point γ at the same instant as the true sun and would meet it again after the lapse of one tropical year.

The mean solar day is divided into twenty-four hours; the hours into minutes and seconds. These are the units indicated on our clocks and watches even though they do not normally indicate the mean time of the meridian.

Therefore, there is a variable difference between true time and mean time during the course of the year: this difference is called the *equation of time* and is defined as the difference between mean time and true time:

$$E = T_m - T_v$$
$$T_m = T_v - E.$$

Table 3 gives a mean value of the equation of time; it is worth noting that it is not exactly equal on the corresponding dates of successive years. The differences, however, are without importance in gnomonics (figure 22).

The true time and the mean time as defined above are counted from 0 to 24 hours, starting from the solar transit at the meridian, i.e., at the moment which we call noon. They are, therefore, respectively equal to the hour-angles of the true sun and of the mean sun.

The mean time, reckoned in this way, is still not practical for our daily life. The difficulty has been bypassed by simply increasing it by twelve hours, that is, by starting to count from midnight. This arrangement is called *civil time*.

Time so defined, true or mean, is still valid only for the meridian through which the true or mean sun goes at noon. This is the *local time*. If we set our watch on the local meridian, the time for Dover, Portsmouth, or Plymouth will be different.

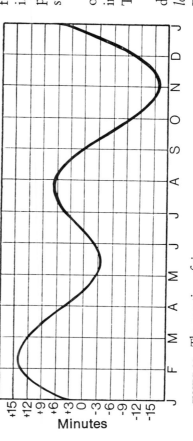

FIGURE 22   The equation of time

This discrepancy was of little importance in the days when travel was very slow. But the difficulties created by the variety of local times became insuperable when time-tables had to be set up for the railroads. A new hour was then introduced, made even more independent of the course of the sun, by defining in the various countries a uniform time over their whole extent. This time is *legal time*. For the British Isles the legal time is the hour of zero meridian, namely Greenwich.

Here is an example of the calculation of the legal time valid on a day when the equation of time would equal —5 minutes 30 seconds, for a longitude of 8° 33′ E, recalling that according to Table 1 this longitude corresponds to 34 minutes 12 seconds, the true time being 14 hours 25 minutes 12 seconds:

$$\text{Legal time} = 14\text{h } 25\text{m } 12\text{s} - 5\text{m } 30\text{s} - 34\text{m } 12\text{s}$$
$$= 13\text{h } 45\text{m } 30\text{s}.$$

Furthermore, what is called *summer time* has been introduced since the First World War for the purpose of saving light and therefore power. Legal time was augmented by one hour in summer in order to increase the length of the afternoon; during the winter this was discontinued because the later sunrise made it unnecessary. In contrast to England, France has kept summer time the year round since 1946 so that the legal time is now (1964) that located on the 15° E meridian, i.e., the meridian going through the neighbourhood of Prague, Czechoslovakia.

It must be admitted that for amateurs in the field of sundials, summer time has the inconvenience of increasing by an hour the difference between the true local time indicated by the sundial and the normal legal time, a difference that is sometimes already considerable.

Legal time, as we have already stated, had to be introduced in every country with the development of the railroads. But the even greater increase in the speed of communications which has taken place since the birth of civil aviation has made the need for unification on a world-wide scale imperative. A new time has been defined, *universal time* (UT), which is based on the Greenwich meridian and is the same all over the world.

Even before the advent of universal time there were efforts to unify the hours, at least for countries located along the same meridian. The meridian was used to divide the surface of the earth into twenty-four time-zones, each having the same legal time. The time zones are numbered from 0 to 23, starting from Greenwich and moving towards the east, and each zone contains 15 degrees of longitude; the first zone spreads from $7\frac{1}{2}°$ W to $7\frac{1}{2}°$ E of Greenwich (meridian 0), the second spreads $7\frac{1}{2}°$ on either side of the 15° meridian and so on. It is impossible to follow strictly the zones delineated by the meridian within a given country, so it has been decided to maintain over the whole of the country the time of the zone into which the bulk of the country falls, if the country considered is not too large. This is the case for England and also for metropolitan France which has adopted the legal time of zone 0 even for its territories located beyond 7° 30′ of longitude. For instance, Strasbourg and part of Alsace belong to zone 1, but they use the time of zone 0. Many legal time zones differing from each other by one hour, depending on the longitudes of the regions, had to be defined for Russia, Canada, and the United States. Russia extends over ten time zones, Canada over six, and the United States over four.

Another difficulty arising from this distribution of time must be considered. Let us suppose that it is 6 o'clock in the morning on Monday in zone 0. It will be 7 o'clock in zone 1, 8 o'clock in zone 2, and finally 5 o'clock Tuesday in zone 23; coming full circle to zone 0, it is still 6 o'clock on Monday. The date then will have changed. A ship going around the world towards the east which advances its clock regularly at each zone crossing, will come to its point of departure one day ahead of its calendar. A conventional *date-changing line* has been established at the 180° meridian in order to avoid this confusion. The eastbound ship will keep the same date for two days, giving to these days the numbers 1 and 2, on crossing the date meridian. A ship travelling in the opposite direction, on the other hand, will have to skip one calendar day on crossing the 180th parallel. The date-change line, although located in the Pacific, does not follow the meridian exactly and takes account of political frontiers in order to avoid different dates in the same archipelago.

8 THE MOON

Moonlight, so celebrated by poets and lovers, has the great advantage of allowing those with a taste for gnomonics to read the approximate time from their dial even during the night. As we shall see, there are certain prerequisites and we must know how to deduce the age of the moon from its appearance, i.e., from its phases.

It is an established fact that the moon is a cold spherical body, like the earth, and that it reflects the light it receives from the sun. It interests astronomers in many ways. The ratio of its size to that of the earth, compared with the moons of the other planets, is very large; this fact, along with other remarkable observations, has led many authors to support the hypothesis that it was separated from the earth in the remote past.

Everyone knows that the moon rotates around the earth in the same direction as the earth itself rotates around the sun, i.e., in the positive direction. Its apparent motion is therefore similar to that of the sun, that is, its meridian transit falls constantly behind the transit of a given star. The daily retardation is much greater than that of the sun because of its proximity, amounting to 54 minutes per day as compared with the stars or, if one prefers, to 50 minutes behind the sun.

For an equal declination of the moon and of the sun, the lunar day proper, if we mean by this the time elapsed between the rising and the setting of the moon, will, therefore, be longer than the solar day defined in the same way.

The observation of the lunar orbit on a map of the sky similar to that shown previously for the sun indicates that the moon draws a great circle on the map and that this circle is not fixed. Besides, it is inclined to the ecliptic by an angle which varies between 5° and 5° 18'. We deduce that the moon's declination may vary between extremes of +23° 27' + 5° 18' and −23° 27' − 5° 18', namely between +28° 45' and −28° 45'. The two intersections of the moon's orbit with the ecliptic have been given the names of *ascending* and *descending nodes*, according to whether the moon rises or descends towards the northern hemisphere.

Let us note a fact which will interest those who happen to look up the great astronomical almanacs. The astronomers found it more convenient to use a system or reference based not on the equator and the celestial poles, but on the ecliptic and its poles, in their calculations for the moon as well as for the sun and the planets. These so-called ecliptic co-ordinates are measured in the same way as the equatorial co-ordinates, in degrees and in hours, still using γ as the origin; but one then speaks of *celestial latitudes* and *longitudes* rather than of declinations and right ascensions.

Various periods of revolution may be chosen for the moon, depending on the point of reference taken for noting the complete revolution. The most important period for us is that which brings the moon to the same position with respect to the earth and the sun. The moon is then in the same phase, called a *synodic month* and

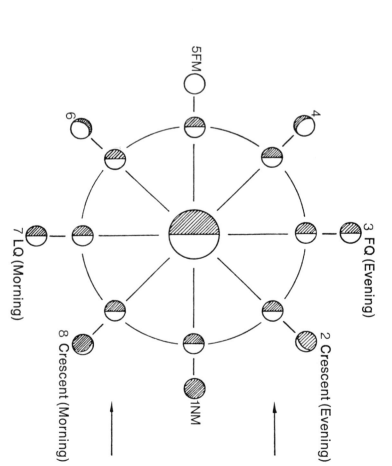

FIGURE 23   The phases of the moon

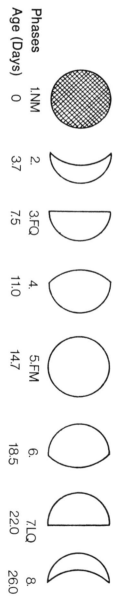

FIGURE 24   The age of the moon
in round numbers for an assumed
lunation of 29·5 days

Phases	1.NM	2.	3.FQ	4.	5.FM	6.	7.LQ	8.
Age (Days)	0	3.7	7.5	11.0	14.7	18.5	22.0	26.0

is defined as the time elapsed between two consecutive full moons or two new moons. This time period, which is also called a *lunar month* or *lunation*, amounts to 29 days 12 hours and 44 minutes, or to approximately 29½ days.

The *phases* of the moon represent a simple phenomenon which is explained by the fact that one of its faces is always lit by the sun and that we see this face from various angles. Figure 23 shows the moon in its successive positions relative to the earth located at the centre, as it appears to an observer on the earth. The terms new and full moon, first and last quarter are known to all. They are usually indicated by the abbreviations NM, FM, FQ, and LQ. The sequence of appearances taken by the moon starting from the full moon will be those shown in figure 24.

It is worthwhile to note the expression "the age of the moon," which is the time elapsed since the last new moon expressed in days. For instance, at the instant of the first quarter, the age of the moon will be about 29.5/4 days, a rounded-off value of seven and a half days in gnomonic practice.

9 DETERMINING THE ORIENTATION OF THE MERIDIAN

The direction of the meridian is determined by finding where the true south or true north lies. Many procedures for this determination are available to gnomonists.

The simplest way (which seldom gives an accurate result) is the use of a good quality compass of sufficiently large dimensions. It is then necessary to know the *magnetic declination* of the location, i.e., the difference in direction between the magnetic north shown by the compass and the geographical north. The declination varies from one place to another; it also varies with time, but slowly.

A simple but more precise procedure for determining the true south consists in installing a gnomon on an accurately horizontal plane. It goes without saying that the gnomon must be vertical and that its installation must be checked with the help of a straight edge. A circle of a given radius is traced around the foot of the gnomon. During the course of the day the shadow of the tip of the gnomon will trace a hyperbola on the plane, which may be marked with a pencil. The intersection of the circle with the hyperbola determines two points A and B which are joined (figure 25). The perpendicular on the middle of AB will necessarily pass through the foot of the gnomon and indicate the direction of the meridian sought.

A double check of the determination by tracing many circles of various radii rather than a single one is recommended. In this fashion many north–south lines will be obtained which should coincide if the observations are done with proper care.

When it is possible to find the longitude of the location on a small-scale map, the following procedure is also possible. A watch set exactly with the time given on the radio or by an observatory clock indicates the legal time, which equals the true local time when corrected for the equation of time and the longitude (expressed in hours), and, if necessary, the added hour of summer time. It is then possible to infer the time indicated by the watch at the instant of the solar meridian transit. If G is the longitude, we have

$$Tm = Tv + E + G(+1h).$$

At the true noon,

$$Tv = 0.$$

We are left with

$$Tm = E + G(+1h).$$

In England, when summer time is in use, the calculation yields, will indicate at this precise instant the exact direction of the north–south axis. Let us work out another example: for a location of longitude 3° 15′ 30″ W and an equation of time on the day of + 10m 45s,

$$\mathrm{T}m = \mathrm{10m\ 45s} + \mathrm{13m\ 02s} + \mathrm{1h} = \mathrm{1h\ 23m\ 47s}$$

$$\mathrm{(or\ 13h\ 23m\ 47s)}.$$

This is the time indicated on the watch at the instant of the true noon. The shadow of a plumb-line on a strictly horizontal plane

$$\mathrm{G} = 6°\ 12′\ \mathrm{E},$$

$$\mathrm{E} = -\mathrm{1m\ 40s},$$

$$\mathrm{T}m = -\mathrm{1m\ 40s} - \mathrm{24m\ 48s} + \mathrm{1h} = \mathrm{0h\ 33m\ 32s}$$

$$\mathrm{(or\ 12h\ 33m\ 32s)}.$$

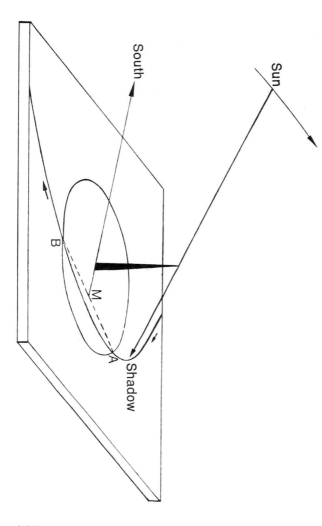

FIGURE 25 The shadow of the gnomon describes a hyperbola which intersects the circle at two points

PLATE 19 One of the dials of Father
Ildéphonse at the convent of
Cimiez-sur-Nice. This dial, 2.45 m across,
has a meridian on each of its hour-lines.
We note their distortion in the reference
system made up of the hour-lines and the
daily arcs of the dial. (*Authorized
reproduction of an illustration in Boursier: 800
Devises de cadrans solaires, Paris, 1936*)

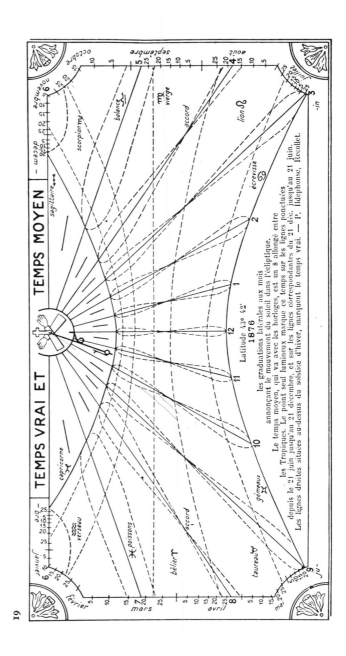

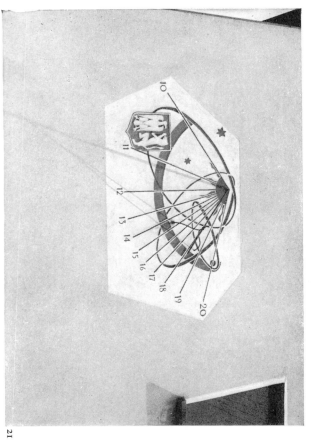

21

20

22

PLATE 20   The network of the Babylonic and Italic hours on the dial of an inn in Gschnitz in the Tyrol hides the dial with Roman numerals on the exterior rim. The area occupied by the dial between the arcs of the solstices takes the shape of a lemniscate.
(*Photograph: Dr Kühnelt, Innsbruck*)

PLATE 21   Recent dial inserted in the stucco of the walls of a school in Bouxwiller, Alsace (*Photograph: the author*)

PLATE 22   The very simple dial of a church in Saverne, Alsace (*Photograph: the author*)

23

PLATE 23   The dial of the Ponte Vecchio
in Florence (*The photograph kindly supplied
by the Ente Provinciale per il Turismo di
Firenze*)

PLATE 24 Two shepherd's dials (eighteenth century). The one on the right is of ivory. They both have a scale giving the elevation of the sun. The hours indicated are those of the morning and evening, respectively. (*Photograph: Strasbourg Museum*)

24

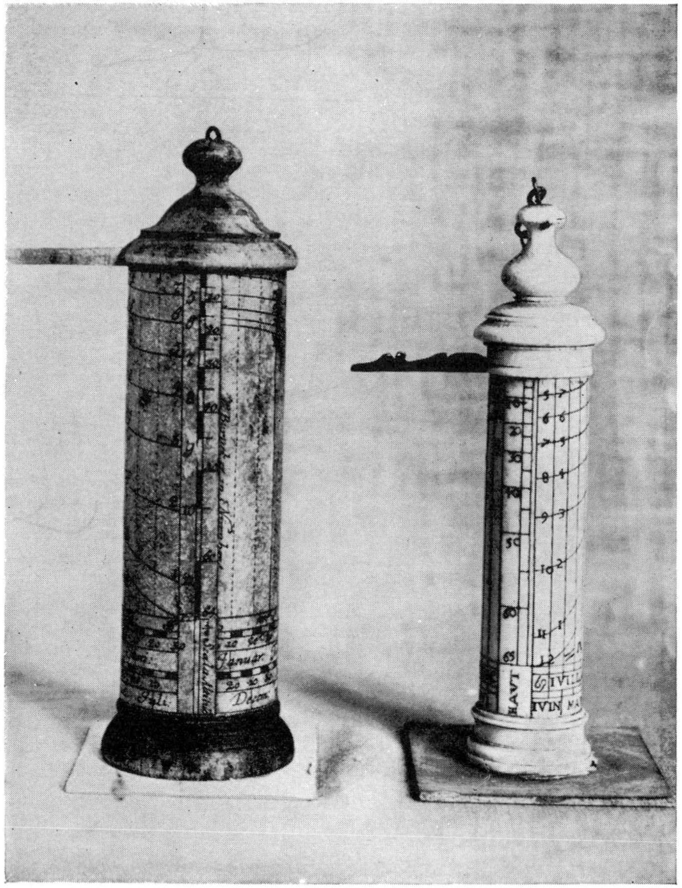

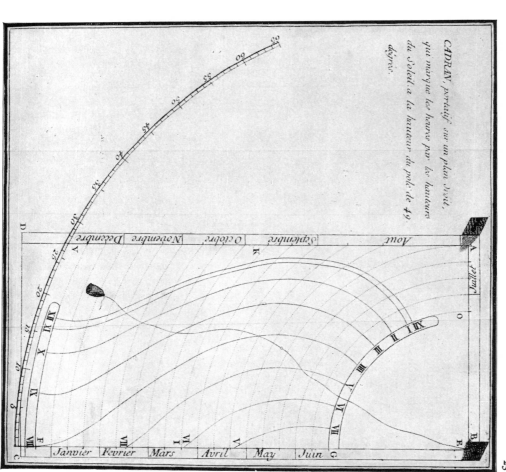

PLATE 25 Old elevation dial eighteenth century, according to an engraving found in the book of Bedos de Celle. The graduated circle at the bottom was erased after the tracing of the curves.

25

CADRAN, portatif sur un plan droit, qui marque les heures par les hauteurs du Soleil. a la hauteur du pole de 49 dégrés.
Juillet
Aout
Septembre
Octobre
Novembre
Decembre
Janvier
Fevrier
Mars
Avril
May
Juin
A
B
C
D
E
F
G
K
O
V

The following procedure, based on the observation of the Polar Star, consists of installing two plumb-lines in such a way that they are in line with the Polar Star (figure 26). The plane formed by the plumb-lines is then the meridian.

This procedure is somewhat inaccurate in high latitudes because the relatively high elevation of the star makes the observation rather difficult. We must remember as well that the celestial pole is not located right on the star but $1°$ away from it, which is equivalent to about twice the apparent diameter of the sun in the direction of the last three stars of the Great Bear, or more precisely in the direction of the star $\in$ of that constellation.

There exist other procedures, current among astronomers and mariners, which require instruments not usually available to the gnomonists; we mention them only for the record.

10 DETERMINING THE ORIENTATION OF A WALL AND THE INCLINATION OF A PLANE

Orientation of a Wall

The vertical plane perpendicular to the meridian at one place is known as the *first vertical*. Its intersection with a horizontal plane – or with the earth's surface where it is flat – is a straight line determining the east–west direction.

In gnomonics, the *orientation of a wall* is defined as the angle which it makes with the first vertical, therefore with the east–west line. This angle was called the *declination* of the wall by the ancient gnomonists. From it we derive the expression declining wall. Used in this way "declination" has nothing to do with the declination of the magnetic needle of a compass or with the

declination of a star, which is its angular elevation above the equator. The triple meaning of the word declination can clearly lead to some confusion, but the text will usually indicate the meaning.

The declination of a wall explicitly indicates the orientation of the wall only when the direction faced by the wall is also stated. In the case of figure 27, the wall whose foundations are indicated by ABCD has two faces: AB and CD. We say then that AB declines towards the SW, CD towards the NE.

(1) The most obvious procedure for determining the declination of a wall is the following: find the direction of the meridian, using one of the means discussed in the previous paragraph, draw it on a strictly horizontal board located along the wall, measure the angle which the wall makes with it, then deduce the declination by adding or subtracting a right angle. Such a measurement is performed with the help of a protractor whose dimensions should be as large as possible.

The declination of the face CD in the above example will be

$$90° - \angle \text{DOS} = \angle \text{DOE}$$

or

$$\angle \text{COS} - 90° = \angle \text{WOC}$$

which is the same thing, and the angle so determined should be followed by the label SW.

(2) If one is familiar with trigonometry and if an almanac giving the declination of the sun for the day of observation is available, one may simply wait for the moment when the sun is in the plane of the wall, i.e., when it no longer, or barely, shines on its rough spots. One needs only then to determine the azimuth of the sun, i.e. the angle formed by its direction with the meridian and to take its complement. The formula to be used is

$$\cot \text{A} \sin \text{HA} = \tan \text{D} \cos \phi - \sin \phi \cos \text{HA},$$

where A is the azimuth of the sun, ϕ the latitude, D the declination of the sun, and HA its hour-angle, which naturally is the true hour-angle, i.e., the angle made by the hour-circle of the sun with the meridian. The angle will therefore

FIGURE 26   The pole lies 1° away from the Polar Star (or twice the apparent diameter of the moon or of the sun) on a straight line which joins this star to the star ε of the Great Bear. When the plane defined by the two plumb-lines passes through this point, it coincides with the meridian

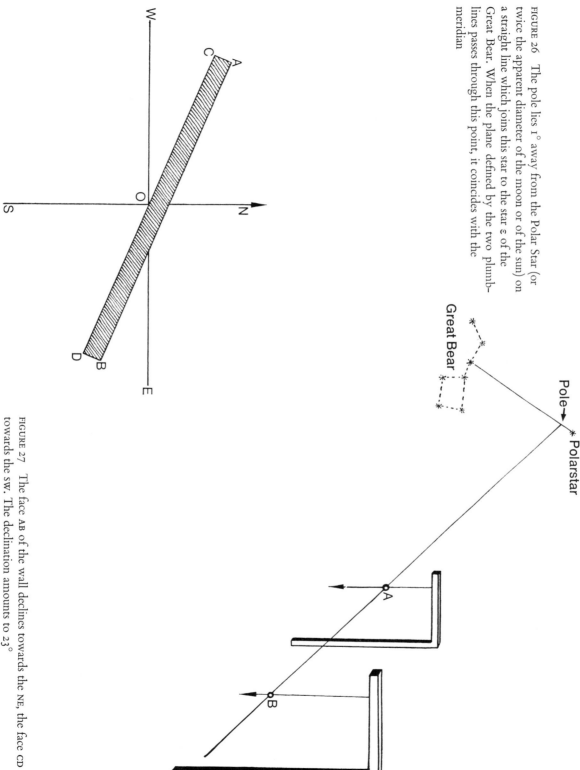

FIGURE 27   The face AB of the wall declines towards the NE, the face CD towards the SW. The declination amounts to 23°

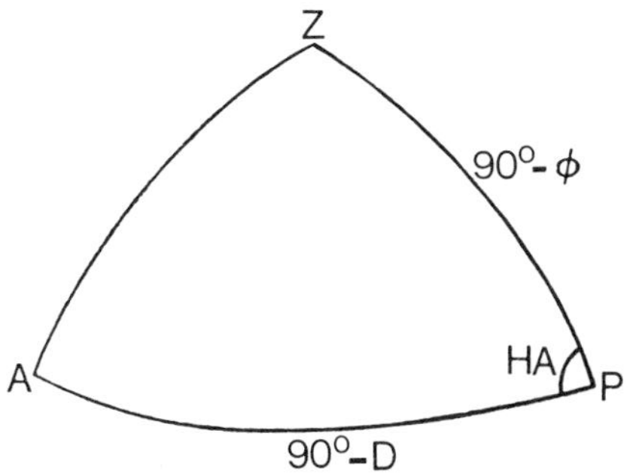

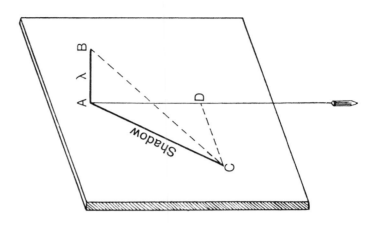

FIGURE 28   The celestial triangle which has as summits the Pole, the Zenith, and a star A

FIGURE 29

be equal to the true time in the afternoon, while in the morning it will be the twelve-hour complement of the true time. We have seen how we may convert from true time to mean time or to legal time. Here we have the reverse. The above formula is one of the relations existing between the elements of the celestial triangle formed by the pole P, the zenith Z, and the star A (figure 28). It is currently used by mariners who need very frequently to measure the azimuths of the sun or the stars to obtain the variation, i.e., the difference between the north shown on their compasses and the true north or the meridian.

(3) An analogous procedure, which possibly gives the most accurate result because it allows an unlimited number of checks, consists in setting against the wall a plywood board on which a perpendicular thin rod AB of length λ has been fixed (figure 29).

A plumb-line is set at the foot A of the rod, which indicates the vertical AD. The shadow of the tip B of the rod AB falls on C on the board. The horizontal distance CD from C to the vertical AD is then carefully measured and, simultaneously, the exact time is noted from a carefully set watch. In figure 30, we then have

$$\tan \angle \text{ABC} = \text{CA}/\text{AB} = \text{AB}/\lambda,$$

which gives the angle ABC with the help of a trigonometric table.

The true time at the moment of observation is then calculated and the azimuth of the sun ∠EBC at that time can be deduced, using the formula given in (2):

$$\angle \text{EBA} = \angle \text{EBC} - \angle \text{ABC}.$$

But ∠EBA = ∠EFB, i.e., this angle is the declination of the wall. Many such measurements, always coupled with careful time checks, should be done: the results will differ slightly, one from another, because of unavoidable errors in the measurements. But their mean will give the declination of the wall to a sufficient degree of approximation.

The Inclination of a Plane

The so-called *inclination* of a plane or a wall in gnomonics is the acute angle made by this plane with the horizontal. Therefore the inclination of a horizontal dial is zero while that of a vertical dial is 90°.

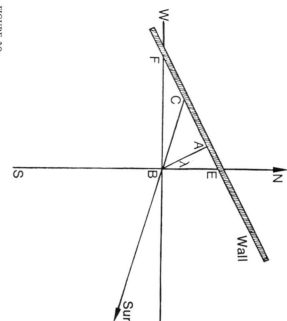

FIGURE 30

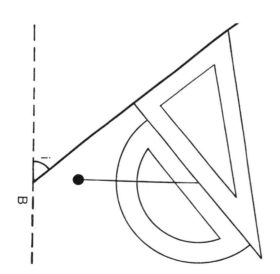
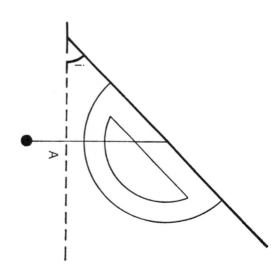

FIGURE 31

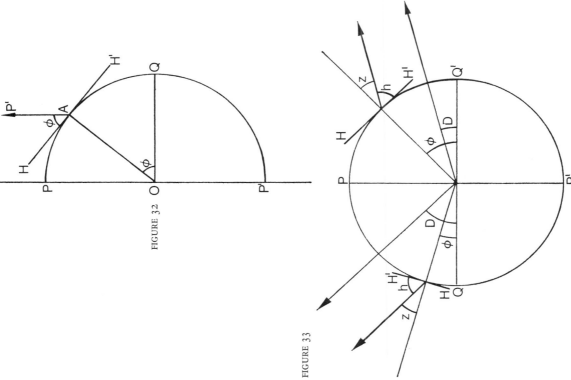

FIGURE 32

FIGURE 33

The determination of the inclination is a simple problem for the gnomonist armed with the three essential instruments, the straight edge, the protractor, and the plumb-line.

Figure 31a shows a plane inclined towards the ground. In this case the angle between the plumb-line and the plane must be measured by laying the protractor along the line of greatest slope. Then the desired angle *i* of the inclination is the complement of the measured angle.

In figure 31b, which represents the case of a plane inclined towards the sky, a straight edge applied along the line of greatest slope is used and the plumb-line is installed as shown. The desired angle *i* is the angle between the plumb-line and the side of the straight edge perpendicular to the plane.

11 THE DETERMINATION OF GEOGRAPHICAL CO-ORDINATES

In this section we give first the principle behind the determination of these co-ordinates, along with a summary of the procedures followed. However, an accurate determination of these quantities requires instruments or installations not usually available to everybody, so we shall confine our description to the practical and really useful procedures that are easily accessible.

We have seen that the latitude was equal to the elevation of the celestial pole above the meridian. In figure 32, OQ is the radius of the earth at the equator while HH′ is the horizon of the observer. If AP′ is drawn parallel to PP′, the angles HAP′ and AOQ are equal, since their sides are mutually perpendicular. It will, therefore, be possible to obtain an approximate value of the latitude by measuring the elevation of the Polar Star above the horizon. Let us recall that all the elevations measured must be correct for atmospheric refraction (Table 2) and that the Polar Star is located at a distance of approximately 1° from the celestial pole. The measurement of the latitude with the help of the Polar Star can be made with precision only when a table of corrections is available from the Ephemeris or a current nautical almanac.

The most precise and widely used procedure consists in measuring the elevation of a star whose declination D is known at the moment of its meridian transit. It follows from the examination of figure 33 that we will have

$$\phi = D - z$$ for the star on the left-hand side,

or

$$\phi = D + z$$ for the star on the right-hand side in the figure,

where z is the zenith distance $90° - h$ of the star, and h is its elevation above the horizon, in accordance with the size of the declination in relation to the latitude. Consequently, to observe the star, we must turn towards either the south or the north.

The determination of the longitude seems simpler, at least at first sight. The difference in longitude between two points A and B, when expressed in hours, minutes, and seconds, equals the time elapsed between the transits of the same star through the meridians of A and B. If we know the time of transit of a star through the Greenwich meridian, the longitude of the location equals the difference between this time and the time of its transit through the local meridian.

The observatories have at their disposal a special instrument called a meridian, a mobile telescope installed permanently in the plane of the meridian: it appears to be perfectly suited to this task. Since the geographical co-ordinates of the observatory are accurately known, the purpose of these meridians is essentially the measurement of the declination and right ascension of the stars. It is sufficient to know the latitude to within half a degree in gnomonics. On the other hand, it is an advantage to know the longitude more accurately: to determine the legal time of a locality, we must add the correction of the equation of time and the longitude to the reading on the sundial.

In gnomonics the easiest way to find the geographical co-ordinates in practice seems to be to look up a small-scale map and abstract them. For instance, a map of 1:50,000 will give a good approximation. If more detailed maps are available, so much the better. We would point out to people in coastal regions the great precision of nautical charts which are always carefully prepared, with the added advantage of a rectangular grid (Mercator's projection).

It is well known that to find the latitude of a point A on a map with a rectangular grid, a line parallel to the geographical parallels is traced through A. The latitude of the particular station can be read on the edge of the map. In order to obtain the longitude, a parallel through A is traced to the nearest meridian and the longitude obtained on the horizontal scale found at the top or bottom of the map (figure 34).

When the map is set on a curvilinear grid as is usually the case with geographical atlases, a parallel and a meridian, which follow as closely as possible the general curvature of the line representing the parallels and meridians on the map, should be traced through A.

The greatest care must be exercised that the scale of longitude found on the map is that based on the Greenwich meridian, the usual scale. In some countries, however, and especially in France, there are still in existence many maps whose origin of longitude is the Paris meridian. We may mention that, while the author was installing a dial, he happened to use one of those excellent tourist-maps issued by the "Club Vosgien" and realized to his astonishment that the origin of longitude on these maps was neither Paris nor Greenwich, but the island of Hierro (Iron Island), one of the Canaries.

Whenever a map set on the Paris meridian is used, it is recommended that the user make a sketch (figure 35) of the meridians of Greenwich and Paris, the one at Paris being located 2° 20′ east of Greenwich, and of the meridian of the point A as well.

Three situations may arise: if A lies east of the meridian of Paris, the longitude shown on the map must be increased by 2° 20′; if A lies west of Greenwich, the longitude on the map must be decreased by 2° 20′; if A lies between the two meridians, the longitude of A equals 2° 20′ minus the longitude shown on the map. In any of these cases, the longitude is east or west depending on whether A is to the east or to the west of the Greenwich meridian.

In the exceptional cases where the meridian of Hierro is the origin, the correction procedure is the same, but account must be

taken of the fact that this meridian lies 17° 40′ west of Greenwich or 20° from Paris.

If no useful map is available, we can nevertheless obtain satisfactory results by using the following procedures:

Latitude

A mariner, with a sextant at his disposal, would know how to use it and there would be no problem. Others must imitate the Ancients and use a gnomon. For this purpose a horizontal board should be laid out; on it a rod of length λ can be used as a gnomon. It is clear that the rod should be perpendicular to the board and

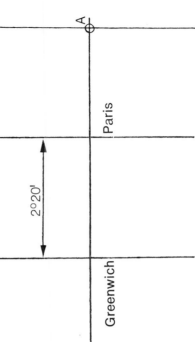

FIGURE 35

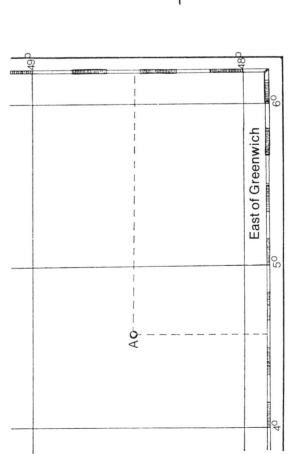

FIGURE 34   The point A lines at 48° 32′ N and 4° 35′ E

this should be verified with the help of a good straight edge (figure 36).

The north-south line is determined first and should be traced on the board through the centre of the foot of the gnomon. At the true noon, the shadow of the rod will reach its minimum length on this line; if this does not happen, the trace of the meridian has been poorly determined.

The instant when the minimum occurs can also be determined by the calculation of the time of the true noon. Let m be the length of the shadow at that instant. The apparent elevation of the sun will be such that $\tan h = \lambda/m$, and h can be checked from a trigonometric table. The true elevation of the sun will be derived from it by applying the correction for the atmospheric refraction (Table 2). The latitude is obtained with the help of the zenith distance of the sun, which is the complement of the corrected elevation. An almanac supplies the declination D of the sun at the moment of observation. Then either of the formulae quoted previously can be used, depending on the circumstances,

$$\phi = D + Z,$$
$$\phi = D - Z.$$

Longitude

The apparently simple problem of the determination of the longitude has been the bugbear of mariners and explorers for centuries because it compares the time of the locality with the time at Greenwich. There is no place in the world today, however, where it is not possible to obtain the time with precision. It is easy to deduce the Greenwich mean time (universal time UT) from the legal time. In addition, we can find the true Greenwich time from

an almanac that gives the equation of time.

We have seen how we can determine the exact time of the meridian transit of a star (figure 17). If this procedure is applied to the sun, it gives the true time of the sun's local meridian transit. The following steps can now be taken to obtain the longitude of a locality.

(1) Infer the Greenwich mean time from the legal time by a simple addition or subtraction of whole hours depending on the time zone. If daylight saving time is in use, the summer hour is to be subtracted from the legal time. The remainder is the mean Greenwich time.

(2) Add or subtract the equation of time at the instant of observation to or from this time. The true Greenwich time, at the meridian $0°$, is then obtained. We may first set a watch to the time given by the radio, then advance or retard it in order to get the

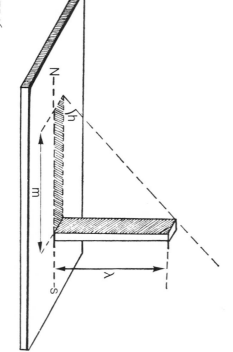

FIGURE 36

mean time of the zero meridian, later repeating either of these actions by an amount equal to the equation of time. The watch will then show the true time at the zero meridian. We go from the mean to the true time, and must, therefore, subtract the equation of time if it is positive and add it if it is negative.

(3) Read the time shown by the watch at the instant of the meridian transit of the sun. At that moment the true local time is noon.

(4) Subtract the time shown by the watch from twelve hours. The difference is the longitude of the locality expressed in hours, minutes, and seconds which may be easily transformed into degrees and minutes.

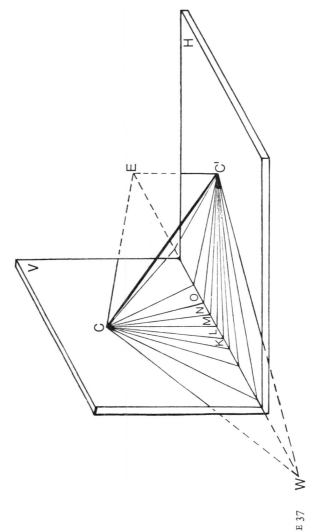

FIGURE 37

CHAPTER THREE

Classical Sundials

I GENERAL INTRODUCTION

Basically, a sundial consists of an object casting a shadow and of a surface on which lines (the so-called *hour-lines*) have been traced; the surface is called the *table* of the dial. The table is most often a plane, but it can be a curved surface: a sphere, a cylinder, etc. Now, however, the latter surfaces are rare. The source of shadow is usually a metallic rod; sometimes it is the edge of a body or perhaps the rim of a surface. As a rule – except for the analemmatic dials – the shadow-casting object is fixed to the table so that the whole instrument forms a rigid and unarticulated piece.

The shadow-casting object may be a small round aperture in a support called a *scaffold* in some old buildings and especially when the installation is meant exclusively to determine the instant of the sun's transit across the meridian. The instrument is then called a *meridian*. The aperture is the *eye of the dial*. In this case, the rays of the sun cast a small white spot on the table which is the image of the sun (figure 4).

For the classical sundials, the style is fixed in such a way that it is parallel to the earth's axis, i.e., to the axis through the earth's poles. It coincides with what is called the *axis of the dial*. For a dial carrying an eye rather than a style, the axis is the imaginary line through the eye which is parallel to the earth's axis.

The *centre of the dial* is the point of intersection of its axis with its table. For a dial with a style, it is simply the point where the style is fixed to the table.

Now let us imagine a horizontal plane H and a vertical plane V, intersecting along the line EW which is oriented exactly in the east-west direction (figure 37). Let us suppose that a straight rod-like object CC' parallel to the earth's axis, is located in the angle formed by the two planes. The rod will then be perpendicular to the line EW. If we draw perpendiculars to this line from C and C' they will meet in M and the plane CMC' will be the plane of the meridian. In other words, the shadow of the line CC' on the two planes V and H will lie along CMC' at true noon. An hour later, the shadow of CC' will be along CNC'. The sun will have covered 1/24th of its daily course during the time interval, or $360°/24 = 15°$, so that the angle between the planes CMC' and CNC' will be $15°$. Similarly at 2 o'clock true time, the shadow COC' will determine a plane which makes an angle of $15°$ with CNC', and so on. Similarly, in the morning, an angular distance of $15°$ will be found between the planes determined by the shadows CMC' at noon, CLC' at 11 o'clock, CKC' at 10 o'clock, etc.

We could determine the true hours in this way by marking off the shadow of CC′ on either plane, from the appearance of the rays of the sun in the morning till their disappearance in the evening. We would then have constructed in the most elementary fashion a double sundial, vertical and meridional on the plane V and horizontal on the plane H, provided that the line EW was properly oriented, the plane H was horizontal, V was vertical, and CC′ lying in the meridian as well as parallel to the earth's axis. The latter condition is satisfied if the angle MC′C equals the latitude of the locality.

In the procedure described above it has been implied that the sun rotates around the axis CC′ rather than around the real axis of the earth; depending on the latitude, the axis CC′ may be removed from the real axis by as much as 6000 km. The hour-angles will not, therefore, be strictly equal because the centre of rotation will

be removed by this distance. However, the great distance of the sun renders the error infinitesimal and completely negligible for our purposes.

It is worth noting that our dial shows the true time. This is not the time that we find on our watches but, as we have seen, it is the basis for all other conventional times. These conventional times can always be deduced rapidly and easily from the true time by following the procedures indicated in the preceding chapter. There is satisfaction in knowing that the true time indicated by our dials will always be correct.

Let us suppose now that while we leave CC′ in its original position, we replace the plane V by a plane V′ which is inclined and not oriented in the east-west direction so that the sundial will be tilted and declining.

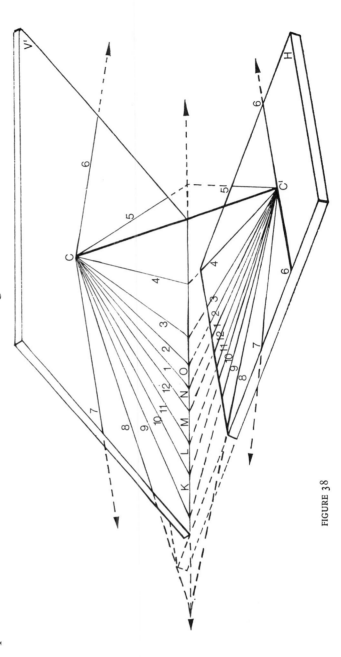

FIGURE 38

Assume that this plane lies beyond the plane H (figure 38). Then the trace of the shadow of cc′ at the various true hours will still be determined by the planes CMC′, CNC′, COC′, etc. Since we have shifted neither the plane H nor the rod cc′, the style of the dial, the centre c of the new dial will be obtained from the intersection of cc′ with v′, the points K, L, M, N, O, etc. by extending to v′ the hour-lines of the dial v′ by simply drawing the lines CK, CL, CM, etc. Here, as before, the shadow will be the intersection of the hour-planes of the sun, i.e., the planes defined by its hour-circles at the various true hours, with the tables of the dials.

Finally, all that has been said still applies if the tables are cylinders, spheres, or any other surface, but the hour-lines would then become curves.

Let us consider finally a flat table whose plane is perpendicular to cc′, i.e., to the style. The plane is then perpendicular to the earth's axis and consequently parallel to the equator. Such a dial is called an *equatorial dial*. This type of dial plays a preponderant role in gnomonics; the following section is devoted to it.

In the preceding it has always been assumed that the northern hemisphere was involved, and this will also be true in what follows. The considerations would, of course, also be true for the southern hemisphere; however, the dial would have to face north to follow the sun and obviously the numbering of the hour-lines would be reversed. Moreover, all that we shall say will apply, by analogy.

2 EQUATORIAL DIALS

An *equatorial sundial* is, as we have just said, a dial whose plane is perpendicular to the style and therefore parallel to the equator. The system of hour-planes described above indicates that the hour-lines of such a dial are straight lines converging at the centre of the dial and making with each of their two neighbours an angle of 15°.

There is no difficulty in constructing such a dial. One has only to draw a circle whose centre is the point of implantation of the style (figure 39), divide it into twenty-four equal parts, and draw as many of the rays of the points of division as are needed; in our middle latitudes the rays should extend from 3 o'clock in the morning till 9 o'clock in the evening since the sun rises between 3 and 4 o'clock in the morning and sets between 8 and 9 o'clock in the evening (true time) on the longest day of the year, June 21. In more elevated northern (or southern) latitudes, more rays will be needed, whereas in lower latitudes their number will shrink to 6 on the equator line.

When a proper site has been found for it, this dial has the advantage (shared, as we shall see, with the horizontal dial) of indicating the time as long as the sun is above the horizon, i.e., during the whole of the day. It represents the image of the plane of the terrestrial equator on a reduced scale, which is, let us recall, the plane of reference used for the measurement of the declinations of the stars and therefore of the sun. The declination of the sun is positive from March 21 to September 23, i.e., during the seasons of fine weather in the northern hemisphere, and the sun will lie above this plane for practically the whole day. It will lie in the extension of the plane during the equinoctial days and it will fall below it during autumn and winter, between September 23 and March 21. In the first case the shadow of the style will rotate on the upper side of the dial, whereas in the last case it will rotate on the lower side, so that this side as well must be equipped with hour-lines but numbered in the reverse direction.

The installation of the dial is as easy as its construction. A protractor is the instrument to use. The style must make an angle with

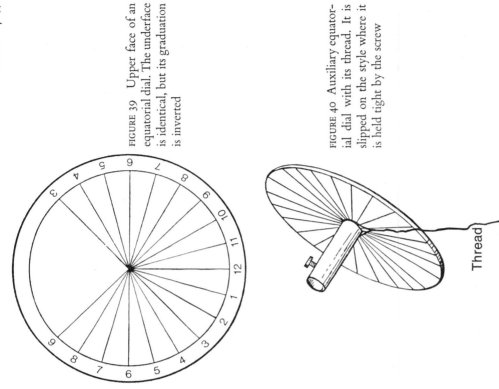

FIGURE 39 Upper face of an equatorial dial. The underface is identical, but its graduation is inverted

FIGURE 40 Auxiliary equatorial dial with its thread. It is slipped on the style where it is held tight by the screw

Thread

the horizontal equal to the latitude. If we take account of the fact that the noon line must be the lowest, or equivalently that the 6 hour and 18 hour lines, which merge into each other and appear as a diameter of the dial, should be horizontal, we need only orient it in the north–south direction to complete the setting.

We have said above that the declination of the sun, i.e., its elevation above the equator and consequently above the plane of the dial, remains constant during the course of one day. This is not strictly true because the variation of the declination is continuous, but over an interval of twenty-four hours the error is practically zero even near the time of the equinox. As a consequence the shadow of the style must have the same length during the interval. The distance between the centre of the dial and the shadow of a point of the style which has been marked, or of the whole style if it does not fall beyond the edge of the dial, should be measured during the course of one day; the next step is to incline and orient the dial by trial and error till the shadow remains the same length during the whole of a diurnal passage of the sun. When this is achieved, the angle of the style with the horizontal equals the latitude and the style points in the plane of the meridian. Such a setting may be effected without any instrument but the dial itself. It is a rather simple procedure for determining to a first order of approximation, but it may serve in many situations.

The equatorial dial thus has the advantages of extreme simplicity of construction and of ease in installation, but, in fact, its usefulness to gnomonics extends far beyond these considerations.

Let us take for example a wall P or any plane surface, vertical, inclined, declining or not, on which the hour-lines of a dial are to be traced and on which there is a properly oriented and solidly fixed style. Suppose that we insert on this style an instrument such as that shown in figure 40, which is a small equatorial dial whose style, exactly perpendicular to its plane, consists of a hollow sleeve which slides perfectly and without play on the style of the dial under construction. The disc of this equatorial dial is turned till the 6h–18h line is exactly horizontal and then the whole is made immobile on the style of the wall by tightening a little screw (figure 41). The loop at the end of the thread is slipped on the stem of the equatorial dial and the thread is pulled in such a way that it stays exactly in the plane of the dial. It is then turned until it coincides with one of the hour-lines of the dial. It is then turned until it coincides with one of the hour-lines of the equatorial dial and is brought into contact with the surface on which the lines are to be drawn; the point of intersection on the surface represents a point of its hour-line. The hour-line itself is easily obtained by joining this point to the centre C.

With this procedure the dial can be drawn reasonably accurately and rapidly. It has been used by craftsmen for centuries for the installation of all types of dials and is especially valuable because it is always applicable, whether the plane of the future dial be horizontal, vertical, declining, or inclined.

The amateur who has no fondness for the mathematical procedures of the *épure* or of calculations will be interested in using it. Its simplicity makes it accessible to everybody. The important thing is to take the necessary time to adjust the sleeve and the style carefully, to see that the sleeve is set exactly perpendicular to the equatorial dial, which must also be accurately drawn, and then to proceed with great care, making sure that the thread remains in the equatorial plane of the dial whose extension it represents.

It is very rare today for anyone to want to draw a dial on a hollow surface like those in the shape of a large shell seen on some of the old cathedrals or on a convex surface or any other three-dimensional surface. For each position of the equatorial dial a series of points along the style can be obtained for each hour-line. Each series of five or six points will thus determine a curve. In this way a complete and accurate dial can be obtained.

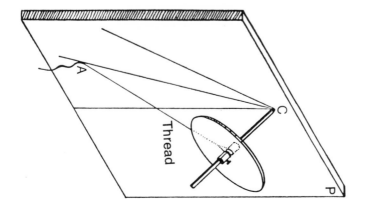

FIGURE 41   The thread goes through the axis of the style and represents the extension of one of the hour-lines of the equatorial. Its intersection in A with the plane of the table of the dial P fixes, with the centre C, the hour-line CA

The equatorial dial is a universal instrument of extreme importance for the practical construction of sundials. Moreover, as we shall soon see, it is also indispensable for the *épures* when they are needed.

The extreme simplicity of construction of equatorial dials has led to a plethora of various, and often strange, creations throughout the history of the sundial; a recent example is the heliograph shown in figure 12, and another the beautiful armillary sphere in plate 41. Recently certain firms specializing in high precision mechanics have made genuine masterpieces to which we shall refer later (for example, see plate 47).

3 HORIZONTAL DIALS

The horizontal dial is the most pleasing in gardens. When set on a stone plate on top of a column and green with the years' accumulation of moss it creates an atmosphere of relaxation and as an ornament rivals many a fine statue.

Next to the equatorial dial, it is the simplest to construct.

Direct Construction

The style is fixed to the table at an inclination equal to the latitude. Dropping a vertical from the style, first a line CM is traced; this is the hour-line of noon. This line must be exactly oriented in the north–south direction when the dial is set in position; the point M should lie in the north and the table should, of course, be strictly horizontal. We then proceed as for an equatorial dial, but here a rigorously flat rectangular piece of plywood P (figure 42) can be used instead of the graduated disc.

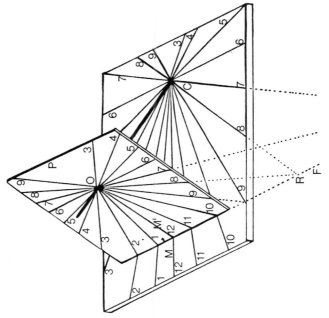

FIGURE 42

A perpendicular is dropped from a point o on the board to M′ on one of its sides. To the left and right of M′, starting from o, the hour-lines of an equatorial dial 15° apart are drawn. The sixth line on either side of OM′ will be perpendicular to OM′; these are the 6 and 18 hour lines which are each the extension of the other.

After these preparations, we drill a hole into the plywood at o, with exactly the diameter of the stem of the style; the board is then slipped onto it as shown in figure 42. The board P is placed perpendicular to the style with the help of a straight edge; care should be taken that all of the side carrying the point M′ lies on the table of the dial. If all these operations have been carried out with the necessary care, the point M′ will fall on the meridian line CM. The extremities of the hour-lines of the equatorial dial will mark an equal number of points on the table, which need only be joined to the point C in order to obtain the corresponding hour-lines of the horizontal dial. If the style is circular in cross-section, its intersection with the table will be an ellipse and the centre of this ellipse would be used as the origin of the hour-lines.

The hour-lines of the equatorial dial whose extremities lie on the lateral sides of the board P will cause some difficulty. This is easily solved by first stretching a thread M′F to represent the extension of the intersection of the board P with the table, and another thread, as described above, from o in the plane of P along one of the hour-lines to its intersection at R with the thread F. With the help of a third thread, held taut and fixed at the centre C, the desired hour-line is determined. This procedure is applied in succession to each of the hour-lines whose extremity is on the left or right-hand side of P. It cannot be applied to the 6 and 18 hour-lines, which are horizontal, but in any case we know that the lines corresponding to them on the dial are perpendicular to CM. The work needs to be done for one side of the dial only; the other side is symmetrical with respect to CM.

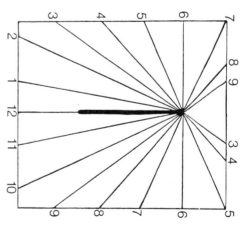

FIGURE 43

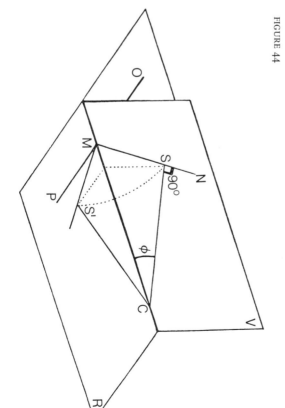

FIGURE 44

Finally, the lines corresponding to the lines of the equatorial dial which are located above the horizontal 6 and 18 hour-lines are obtained by extending beyond c the lines already traced, which are removed from them by twelve hours, and which are their extension through o on the equatorial dial (figure 43).

Construction with the Help of an Épure

The construction of a horizontal dial with the help of an *épure* is simply a repetition of the previous procedure except that drafting tools are used. Attention and imagination are requisites for the amateur, at least in the initial stages, but those who have some elementary knowledge of descriptive geometry will find the process simple. Its main advantage lies in its almost automatic accuracy.

Let R be the rectangle which will serve as the table of the dial and on which the point c will be the centre. As previously, we draw the 12 o'clock hour-line CM, which will be oriented later in the plane of the meridian. The style CS, set in C, will be positioned in the plane v, perpendicular to the plane R along CM (figure 44). The plane of the auxiliary equatorial dial will be a plane perpendicular to SC through S, which cuts the plane v along the straight line MN, perpendicular to SC, and the plane R along OP, perpendicular to CM.

In order to obtain the point M so defined, we draw the triangle CMS, right-angled at S. Suppose that it is mapped onto the plane R using CM as a hinge; the projection allows us to draw the triangle in the plane R (figure 44). S' representing the mapping of S and MS'C the mapping of the triangle MSC. Of course, the angle MSC', as well as the angle MCS, is equal to the latitude and the point M is immediately obtained by drawing a perpendicular through S' to CS'. OP is perpendicular to CM at M, being the intersection

of the plane of the auxiliary equatorial dial with R. The dial will be traced in the plane R by a mapping of the plane of the equatorial dial. Just as the plane CMS could have been projected equally well towards the right or towards the left, the plane of the equatorial dial can be projected above or below. If it is projected below, the image of the circle drawn by S around M will be a circle having MS' for the radius and M for the centre (figure 45). It cuts CM in S', which is the projection of S in the plane R. Now we may trace the circle of the equatorial dial with S" at its centre in the plane R. The radius of the circle should be chosen as large as possible for increased accuracy. The image of the equatorial dial is obtained by dividing it into twenty-four equal parts. It is only necessary to extend its hour-lines until they intersect OP, the intersection being the points of the corresponding hour-lines of the horizontal dial.

The difficulty encountered in the practical procedure again arises; i.e., the hours after 5 o'clock fall beyond the edge of the drawing in figure 45. In that case a simple calculation, which will be applied to the point F for the sake of illustration, will suffice.

If H and G are the intersections of the hour-lines CF and S'F with the edge of the drawing, which must be parallel to CM, we have

$$\text{PH}/\text{PG} = \text{MC}/\text{MS}''$$

or

$$\text{PH} = (\text{MC}/\text{MS}'') \cdot \text{PG}.$$

The lengths MC, MS", and PG may be measured, and therefore PH can be calculated.

Construction using Mathematics

We may obtain the values of the angles (such as MCH) between the hour lines and the noon line (figure 45).

Considering our horizontal dial in perspective, we simplify it to a single hour-angle CA and the style CS in addition to the noon line CM (figure 46). The plane SCM, the plane of the meridian, is perpendicular to the plane MCA. The angle SCM is the latitude ϕ. Let SM be perpendicular to CS and let the plane SMA be perpendicular to SC. The angle MSA is the hour-angle of the sun and MCA, the angle z, between the hour-line CA and the noon line corresponding to the given hour-angle HA, i.e., the angular distance between the true time and noon, is the angle we wish to calculate. We have

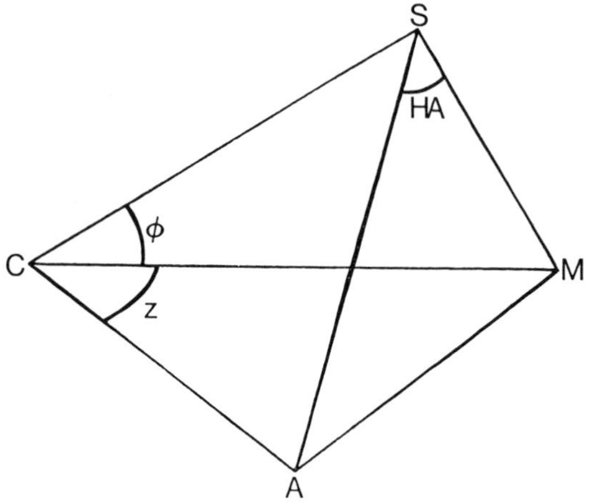

SM/CM = sin ϕ,
CM/AM = cot z,
AM/SM = tan HA.

Multiplying the three equations, we obtain

1 = sin ϕ cot z tan HA,

or

tan z = sin ϕ tan HA.

We find the values of z for each hour by giving HA the values of 15°, 30, 45°, etc. corresponding respectively to 1 o'clock, 2 o'clock, 3 o'clock, etc. The angles with the line of true noon for non-integral hours (half-hours, etc.) can be found with the same case.

This calculation provides a simple example of the derivation of trigonometric formulae for the calculation of the angles of the hour-lines. Similar proofs will not be given for the other dials to be studied because a completely general formula for the calculation of these angles which holds for all sundials, will be derived at the end of chapter four and applied to particular cases.

The style is exposed to all sorts of dangers when the sundial is installed in a garden. A ball thrown by children, if at all heavy, may alter its setting. It is, therefore, customary to replace it by a metallic piece of the shape shown in figure 47.

Allowance must be made for the thickness of this piece in the construction of the dial: the shadow-casting edge will be the upper western edge in the morning and the eastern edge in the afternoon. The distance between the two halves of the dial should be equal to the thickness of the plate-style as shown in figure 48, where the thickness of the plate is purposely exaggerated.

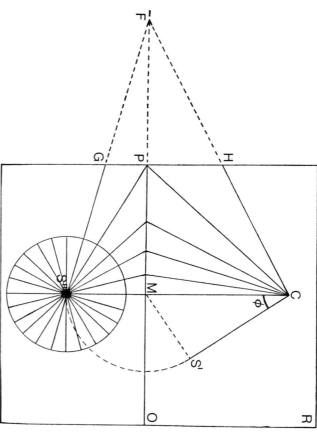

FIGURE 45

FIGURE 46

Let us state once and for all that this procedure is valid for all dials, whatever their orientation in space.

4 VERTICAL DIALS

General Remarks

As we have seen, the table of an equatorial dial must always be perpendicular to the direction of the style, while the table of the horizontal dial must remain horizontal. No other choice is possible in the definitive installation of dials of these two types.

Vertical dials, on the other hand, may have all possible orientations. In the exceptional cases when the table of a dial is located in the first vertical, i.e., in the east–west line, we speak of *non-declining dials*: their table faces the true south or north. When the dial is oriented exactly towards the south, it is said to be *meridional*; when it faces north, it is called *septentrional*. There is also a distinction between vertical dials that are facing exactly the east or the west, i.e., whose declination is 90°: they are called *oriental* and *occidental*, respectively. In general, the term *declining dials* is used for all other orientations.

There are some general facts that still apply to the construction of vertical dials. The noon line is still the intersection of the vertical plane of the table of the dial with the plane of the meridian, also vertical, which contains the style. It follows that the noon line (or the twelve–hour line) of a vertical dial is always vertical.

We have also shown how an equatorial dial could be used to determine the hour-lines of any dial. The same procedure thus applies to vertical dials. If a horizontal dial built for the latitude in which we work were available, we could apply the method described on p. 46 to trace the same hour-lines. These are all

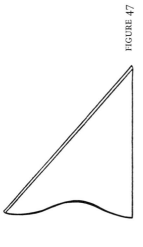

FIGURE 47

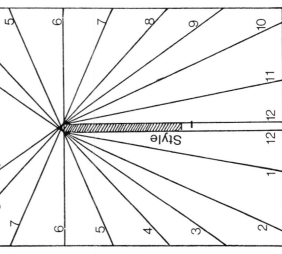

FIGURE 48   When the style is made up of a substance which has a definite thickness, the dial must be divided into halves and these must be placed on either side of the style. The dial then has two centres

simple procedures which are always applicable, but for worth-while results they should be executed with great care.

Let us now review the procedures for drawing the various vertical dials with the help of an *épure*. The pertinent trigono-metric formulae for the direct calculation of the hour-lines will be given later for individual cases. The formula established in chapter four is complicated, but it leads to all the other much simpler formulae pertaining to the various particular cases.

Non-declining Vertical Dials

An auxiliary equatorial dial such as that used in drawing the horizontal dial (see p. 51) is used.

Let v be the table of our dial (figure 49), c the point of im-plantation of the style, cs' the projection of the style on the table. We know that it lies in the vertical plane through cs. If we wish to use an auxiliary equatorial dial, it should be in a plane per-pendicular to the style, and the intersection of this plane with the table of the dial must be determined.

This procedure is the same as that used for the horizontal dial, the plane of the meridian being projected on the table using cs' as a hinge. The tip of the style, whose projection is s', falls on s so that the angle s'cs equals the complement of the angle of latitude. The plane of the equatorial dial, perpendicular to the style, has its trace along sM, the angle csM being by necessity a right angle. The point M is the intersection of cs' with the perpendicular to cs at s and is also a point of intersection of the auxiliary equatorial dial with the table. This intersection is perpendicular to the projection of the style cs', and is, therefore, the straight line op perpendicular to cs' at M.

As for the horizontal dial, the equatorial dial may be projected on the table towards the upper or lower side. We have projected

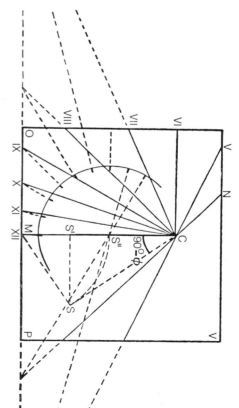

FIGURE 49 To make the *épure* simpler, only the left side of the dial has been drawn. The right side shows only the projection of the style and the two hour-lines which give, by extension, those of 4 and 5 o'clock in the morning. The auxiliary equatorial dial has been projected upwards. The straight line op is simultaneously the equatorial and the inferior edge of the dial

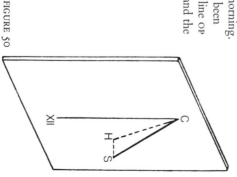

FIGURE 50

it upwards in our diagram, tracing a circle of radius MS, which cuts CS' in S". We now draw around S" the projection of the equatorial dial, using an arbitrary radius, but one that is as large as possible. Since S"M is an hour-line of the auxiliary dial, the other hour-lines are traced by keeping them 15° apart. Their points of intersection with OP determine the hour-lines of the dial being built. For those hour-lines whose intersection with OP falls beyond the table, ratios are calculated by the method described for the horizontal dial (see p. 51).

In non-declining vertical dials, the angle between the hour-lines and the vertical twelve-hour line is given in general by the formula

$$\tan z = \cos \phi \tan \text{HA},$$

where z is the angle made by the hour-line with the vertical twelve-hour line, ϕ the latitude, and HA the hour-angle of the sun, i.e., the number of hours separating the hour-line to be drawn from true noon.

Declining Dials

Installation of the Style

On p. 48 it was pointed out that the practical procedure of inserting an auxiliary equatorial dial on the style was valid for all dials, whatever their surface or orientation. It is, therefore, the best procedure for the direct tracing of the hour-lines in the construction of declining vertical dials, by far the commonest of all dials now in use.

There is, however, an even simpler procedure – just set a good watch to the true local time and mark on the table of the dial the lines of shadow thrown by the style on the hours.

This is a general procedure for all dials whose style is inclined. In the case of vertical dials, it is true noon when the shadow of the style falls on the vertical to the centre, i.e., the hour-line of noon. It is advantageous to apply this method only when the equation of time is zero. The results, it must be understood, will, of course, be good only if certain conditions are satisfied: (1) the wall must be completely flat; (2) the style must be in its proper position; (3) the setting of the watch must be accurate and the true time properly calculated.

Here we must digress in order to study the means of overcoming the difficulties that arise when trying to orient the style properly on a declining dial.

The principle is simple. The style lies in the plane of the meridian and its axis makes with the vertical of noon an angle equal to the complement of the latitude. By using this fact and exercising care, we may put it in its proper position.

We shall now define some of the terms to be used in what follows. Let us consider a dial in which CS is the style and H the projection of the tip of the style on the table (figure 50). The orthogonal projection CH of CS is called the *substyle*.

The height HS of the tip above the plane is called the *straight style*, H being the *foot of the style* and C the centre of the dial.

It is easy to draw the substyle. Let us assume that we draw on the dial under consideration a horizontal plane containing the straight style HS (figure 51). If M is the intersection of the noon line (the vertical through C) with the plane, then the angle MCS equals $90° - \phi$, the complement of the latitude. CH is the projection of CS on the vertical plane, and MS its projection on the horizontal plane. Let us rotate the horizontal plane around the intersection xy of the two planes so that they lie in the same direction (figure 52). If we trace a circle of radius MS, with centre M, cutting xy in S', we have the image of the true length of the style in CS' and MCS' is the angle it makes with the vertical CM, i.e., $90° - \phi$. On the other hand, the angle SMS' is the complement of the declination of the plane of the dial, because MS is the trace of the meridian, namely the direction of true south.

We may then proceed with the very simple construction outlined below.

On a horizontal line *xy* the angle MS′C equal to the latitude is constructed, S′C being equal in length to the length of the style. The perpendicular CM is dropped from C on to *xy* and the angle SMS′ made equal to the complement of the declination of the plane of the dial. An arc of a circle with centre M, passing through S′ gives the point S, and the perpendicular SH to *xy* yields the point H. The straight line CH is then the desired substyle.

The length of the substyle may also be obtained with the help of mathematics. If λ is the length of the style in figure 51, we have

$CM = \lambda \sin \phi$,

$MS = \lambda \cos \phi$,

$MH = MS \cos (90° - d) = \lambda \cos \phi \sin d$.

Once CM and MH are known, it is easy to draw the substyle. However, it is

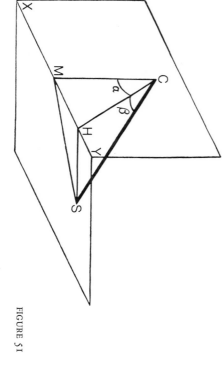

FIGURE 51

always worthwhile to check the accuracy of the drawing by calculating the length of the substyle:

$$CH = \sqrt{CM^2 + MH^2} = \lambda\sqrt{\sin^2 \phi + \cos^2 \phi \sin^2 d}$$

$$= \lambda\sqrt{1 - \cos^2 \phi \cos^2 d}.$$

The angle α which the substyle makes with the vertical CM can be deduced from

$$\tan \alpha = \frac{MH}{CM} = \frac{\lambda \cos \phi \sin d}{\lambda \sin \phi} = \frac{\sin d}{\tan \phi}.$$

We also obtain the height HS of the tip of the style above the plane of the dial from the formula

$$HS = MS \sin (90° - d) = \lambda \cos \phi \cos d.$$

On the other hand, the angle β between the style and the substyle is obtained from

$$\sin \beta = HS/\lambda = \cos \phi \cos d.$$

All these results may prove very useful as a last check during the installation of a style.

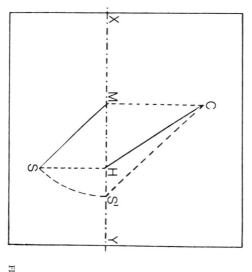

FIGURE 52

Let us now return to figure 52, which provides us with the elements necessary for the installation of a style. Given the point c and the hour-line of noon CM, we add the angle MCH measured with the help of a protractor or by applying the triangle CMH to the wall. We thus obtain the substyle CH. A carpenter's square is applied at H to represent a straight style on which a height HS is measured, thus giving the point s through which the style must pass.

We consider this method of installing the dial the most precise as well as the most convenient.

A number of methods described in old treatises appear simpler but they are never as precise. For example, one may cut out of a piece of rigid cardboard or a thin sheet of plywood (figure 51) the triangle CMS, which has a right angle at M and whose dimensions are supplied by the construction of figure 52. It is applied against the wall along the line CM, the point C of the triangle coinciding with the end C of the style to be installed. At the instant of true noon, determined with the help of a watch set on local true time,

the triangle is turned around CM in such a way that the shadow of CS on the wall coincides with the hour-line CM. The position of the side CS of the triangle represents the position the style should occupy.

If the dial is strongly declining, i.e., if its declination is almost a right angle, it may happen that its centre falls outside the table and even becomes inaccessible. In this case again the construction is simple if the reader is familiar with trigonometry. As previously, he calculates the angle between the substyle with the vertical and that between the style with the substyle. He determines afterwards, by trial and error, according to the dimensions of the dial to be constructed, the length of the style needed to find the height HS of the tip of the style above the table.

These dials have only partial styles (figure 53), which are fixed with the help of two supports.

If we wish to avoid mathematics, we still have to set a watch temporarily on the true local time. We then have recourse to a straight style HS (figure 54) which is installed on the table in such a

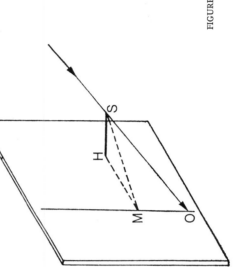

FIGURE 53

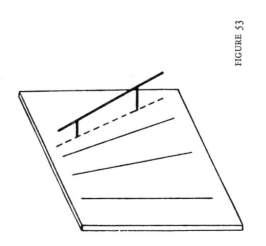

FIGURE 54

way that the point s occupies as exactly as possible the point above the table of the dial where the extremity of the style will lie later. At the true noon, the shadow of the point s determines a point o on the table, which belongs to the vertical line of noon. This line may then be traced. The horizontal HM is traced in pencil. The exact length of MS is measured and, following the method used by the Ancients, a trapezoid ABCD in which DC equals MS, the angle in D being a right angle and the angle C being equal to the latitude, is cut out on a piece of cardboard. The length of AD is arbitrary in principle but it should be chosen as large as possible, depending on the dimensions of the dial.

The cardboard is applied against the dial in such a way that the point D falls in M, the point C in S, and the side DA coincides with the noon line. Under these conditions, the side BC of the card-board represents the position of the style's axis.

Let us mention that once the straight style is installed (figure 54), the position of the point M and of the hour-line of noon whose position it determines, may be obtained in the following way: the construction is brought into the plane of the table by rotating it around a horizontal line going through H and then the side HS and the angle HMS (equal to 90° − d) are obtained from the right-angled triangle HMS.

We note that during the installation of this type of dial the measured length MS and consequently the base DC of the trapezoid will generally be slightly shorter than the width of the dial.

Construction of the Declining Dials whose Centre is Accessible

A WITH THE HELP OF AN AUXILIARY EQUATORIAL DIAL When dis-cussing equatorial dials, we pointed out how any dial could be drawn once the style was properly placed; this, of course, holds

good for declining dials as well. We have also mentioned that this procedure has a very long history and was used by the craftsmen of former times. Since it is a general procedure it will not be re-viewed here, but we recommend it to the tinkers who work with the necessary care.

B WITH THE HELP OF AN AUXILIARY HORIZONTAL DIAL This pro-cedure assumes that a horizontal dial has been built according to the instructions already given. It is then necessary only to trace the auxiliary dial on a drawing sheet.

Suppose then that we have our vertical dial v, c being its centre, cs the style, and cm the noon line (figure 56). We set a horizontal shelf T along the inferior side xy, on which the horizontal dial is placed in such a way that the extension of the style cs passes through the centre c′ of the horizontal dial. The point c′ is fixed on the table with a pin and the horizontal dial rotated until its noon line passes through M. By extending the hour-lines, we ob-tain, from their intersection with the line xy, points on the cor-responding hour-lines of the vertical dial. These we need simply join to the centre c.

It is also possible to use the diagram method assuming that the plane T of the shelf has been rotated downwards around xy. If c is the centre of the dial being built and cm, its noon hour-line, the projection of the plane of the meridian (and therefore of the style) on that plane is obtained by drawing the angle d of the declina-tion and by raising the perpendicular in m to its open side (figure 57). In c, we construct the angle 90° − φ along cm (the comple-ment of the latitude) which cuts xy in n. An arc of circle of radius nn with centre n gives the intersection c of the axis of the style with the projected horizontal plane. All we now need to do is superimpose on the drawing the outline of the horizontal dial (traced right into the *épure*) in such a way that its centre falls in c

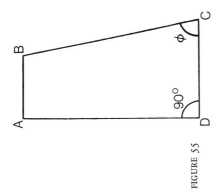

FIGURE 55

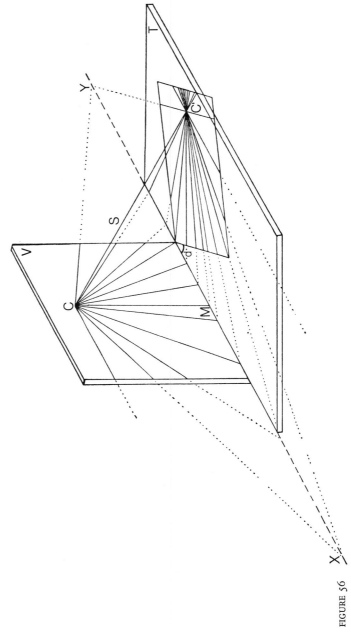

FIGURE 56

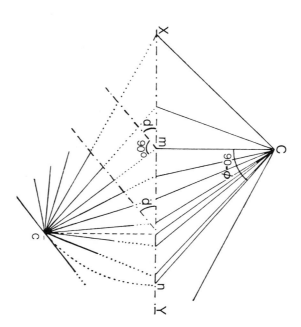

FIGURE 57

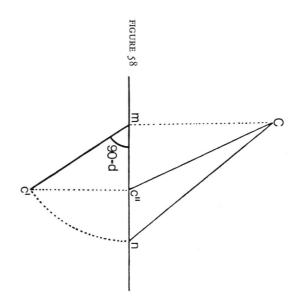

FIGURE 58

and its line of noon along *cm*. To complete the dial, we join the intersections of the extensions of the hour-lines of the horizontal dial with *xy*, which give the corresponding points of the hour-lines of the vertical dial, to the centre C.

C SKETCHING THE ÉPURE BY USING AN EQUATORIAL DIAL Since declining vertical dials are by far the commonest, we now outline step by step the sketching of the *épure* in rather more detail so that it can be used by readers whose knowledge of descriptive geometry has become blurred or even by those who have no knowledge of it at all.

Let us assume that we have the table with the centre of the style and that the projection of the latter has been drawn exactly according to the procedure just outlined. This brings us to figure 58.

Let us visualize the drawing in three dimensions. The style and the substyle (the orthogonal projection of the table) lie in a plane perpendicular to the plane of the dial (figure 59). We wish to place at the tip s of the style a plane perpendicular to it to support the auxiliary equatorial dial. Basically the intersection SE of the plane with the plane CSS' defined by the style and the substyle is all that is required. Since we cannot draw in three dimensions, we rotate the plane CSS' around the straight line CS', bringing it on to the plane of the dial exactly as if we were turning the page of a book. The perpendicular S'S, itself perpendicular to CS' and the length S'S" equals the length of *c"c'* in figure 58. In figure 59, the plane of the equatorial dial is still perpendicular to CS", having been carried around by the projection. It cuts the projected plane along a straight line S"E which cuts CS' in E. In this way, we obtain the point E which marks the intersection of the auxiliary equatorial dial with the plane of the dial. But the plane of the equatorial dial is perpendicular to CS; therefore its tracing on the plane of the dial is perpendicular to the projection CS' of the style. This is arrived at by drawing a straight line perpendicular to CS' in E.

FIGURE 59

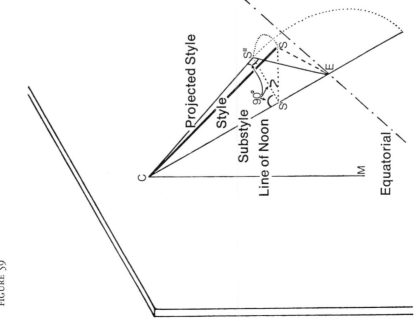

The straight line which is the intersection of the auxiliary equatorial dial (perpendicular to the style) with the plane of the dial is called an *equatorial*.

Figure 60A shows, in three dimensions, the installation at this stage, consisting of the style, the equatorial dial, and the equatorial. From our previous discussions, the reader knows why we refer to the auxiliary equatorial dial and how we intend to use it. We wish to rotate it around its centre in such a way that its noon line will eventually meet the intersection M of the equatorial with the noon line of our prospective dial. Again this is a drawing which cannot be done in three dimensions. Therefore, once again a projection is made. This time we turn the auxiliary equatorial dial around the equatorial (figure 60). The point S then rotates around E (figure 59), using SE as a radius, and falls on the extension of CE. We now use the results already obtained in figure 60B. Since ES in three dimensions is equal to ES'', we trace an arc of a circle around E going through S''. The point S_H, the projected centre of the auxiliary, is obtained in this way; we join it to the point M of the equatorial; this yields the corresponding points of our declining dial which we need only join to the centre C in order to complete the drawing. Figure 60B shows the complete *épure* in which, however, for the sake of clarity, we have put only the noon hour-line and the one o'clock hour-line.

D SKETCHING THE HOUR-LINES OF DECLINING DIALS WITH THE HELP OF MATHE-MATICS Chapter four gives the hour angle z from the value of its tangent:

$$\tan z = \cos \phi / (\cos d \cot \text{HA} + \sin d \sin \phi),$$

where ϕ is the latitude, d the declination, and HA the hour-angle (positive or negative depending on whether it is morning or afternoon).

The general formulae of the same chapter also supply the angle β made by the substyle and the noon hour-line:

$$\tan \beta = \sin d \cot \phi.$$

They supply too the value of angle α that the style makes with the substyle by:

$$\sin \alpha = \cos \phi \cos d.$$

Finally the hour-angle γ for the instant when the shadow of the style falls on the substyle is given by

$$\tan \gamma = \tan d / \sin \phi.$$

Construction of the Declining Sundials whose Centre is Inaccessible

A THE PRINCIPLE Once the style is placed we use the methods just described, which give a point for each hour-line on the table of the dial. But the centre is now outside the table, sometimes even very far removed, so that it cannot be used to provide the second point of the lines to which we could have joined the points just obtained.

There remains the solution of repeating the same method of construction for another position of the style, which for the sake of accuracy is chosen as far as possible from the first location. The second series of points belonging to the different hour-lines so obtained will provide the means for tracing these lines by joining the points belonging to the same hour-lines. Whatever procedure is followed, it is clear that the hour-lines, once drawn, will be arranged in such a way that they converge towards the inaccessible point C of the vertical dial.

A new difficulty arises from the fact that the inaccessible centre of the dial cannot be used to obtain the noon line. Either of the following methods will circumvent this difficulty:

(1) A small *épure* (figure 61) containing the right-angled triangle M'C'H in the horizontal plane going through the support HC' of the style is made; M' is a vertical point of the centre, i.e., of the noon line, the angle $90° - d$ being the complement of the declination of the dial.

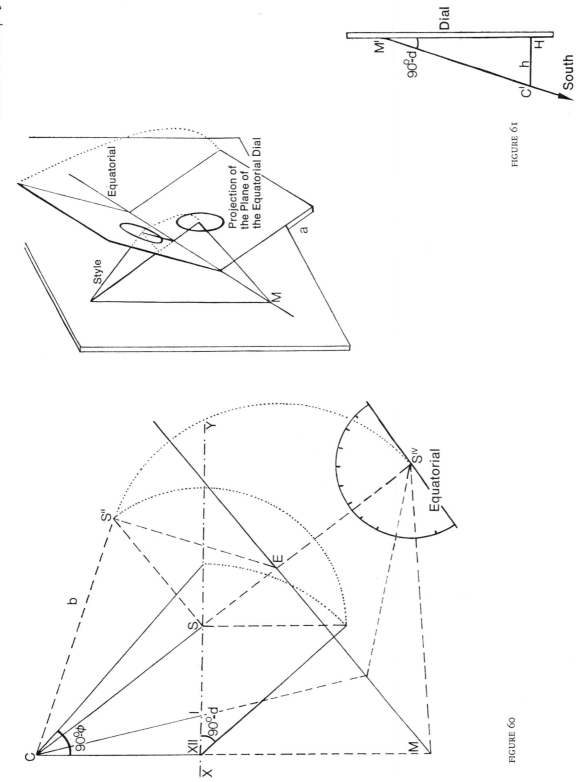

FIGURE 60

FIGURE 61

In this way the distance нм′ of the vertical noon line of the point н is obtained.

(2) This same distance ᴅ is calculated with the help of the formula:

$$\text{м′н} = c'н/(\tan(90° - d)) \quad \text{or} \quad \text{ᴅ} = h \tan d.$$

ʙ CONSTRUCTION WITH THE HELP OF AN AUXILIARY EQUATORIAL DIAL. The auxiliary dial is slipped along the style and brought to a point c′ as low as possible; as usual, we ensure that its plane is perpendicular to the style and we rotate it around its centre (i.e., around the style) until the extension of one of its hour-lines goes through the noon hour-line of the proposed dial. The intersection of the plane of the auxiliary equatorial dial is an equatorial and the intersection of the extension of the other hour-lines of the equatorial dial with the equatorial supplies as many points as there are hour-lines to be traced (figure 62).

The same operation is repeated in c″, chosen as high as possible, where it will provide a new set of points. The desired hour-lines of the dial to be drawn are obtained by joining these points with those supplied by the first operation.

Finally, let us mention another rather laborious procedure for making this type of dial which permits us to obtain the hour-lines from the sun itself. This procedure is especially appropriate when the dial has somewhat large dimensions.

We assume a fictitious style on the dial, namely a rod perpendicular to the table which must be located on a point of the substyle and whose length must equal the height of the corresponding point of the axis of the style at that point (figure 63). An accurate watch is set temporarily on the true local time. At the hours, the shadows ᴋ₁, ᴋ₂, etc. are inscribed on the table of the hours, the shadows $κ_1$, $κ_2$, etc. are inscribed on the table of the tip s of the fictitious style on a day ᴅ₁, preferably close to the solstice. The shadow for only one of these hours has been shown on the figure for the sake of clarity. The same operation is re-

FIGURE 63

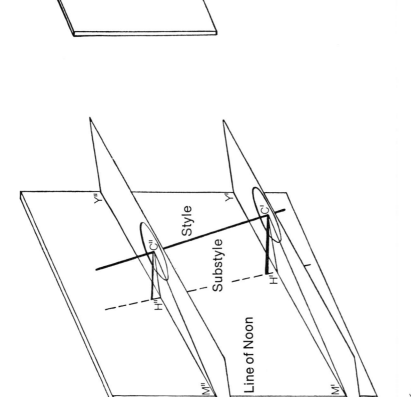

FIGURE 62

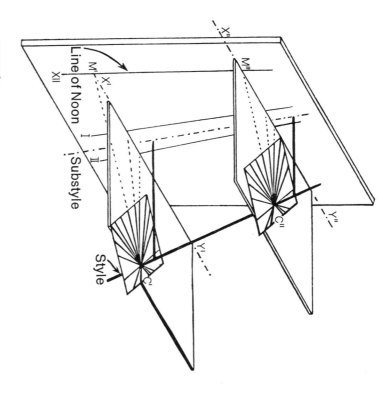

FIGURE 64

peated on another day, about six months later, and therefore again close to the solstice. Of course we obtain different points K_1', K_2', etc., since the declination of the sun has changed by approximately 47°. By joining K_1K_1', K_2K_2', etc. we arrive at the hour-lines with a degree of accuracy that depends on the size of the dial.

C CONSTRUCTION WITH THE HELP OF AN AUXILIARY HORIZONTAL DIAL. The auxiliary horizontal dial is applied on a perfectly horizontal shelf against the dial, exactly as in the previous case when the centre was accessible, so that the style passes through the centre c' of the auxiliary dial (figure 64) and the noon line of the latter meets the noon line in M'. The intersection of the other hour-lines with the horizontal $x'y'$ (the line of intersection of the two planes) yields a set of points belonging to the proposed hour-lines. The same operation is repeated at the point c'' of the style and yields another set of points which lie closer to one another. These points are joined to the corresponding points of the first set.

D CONSTRUCTION WITH THE HELP OF AN *épure*. If an *épure* is used, the whole procedure described on p. 61 is repeated for the two points $c'c''$ of the style, and the points obtained are joined. It is wise to make a preliminary sketch of the drawing to get an idea of how the dial will look. This should make it possible to place the dial on the table.

Let M' and M'' be two points on the noon line, taken in the lower and upper portions of the dial respectively (figure 65). Perpendiculars are made from M' and M'' to $M'M''$ and the complement of the dial's declination is drawn twice, as shown in the figure. From an arbitrary point c_1' we draw the perpendicular to $M'c_1$, which meets one side of the angle of declination in c_3'. The reader will realize that the drawing represents the projection by rotation in $M'c_1'$. An arc of circle of radius $M'c_3'$ is traced around M' giving

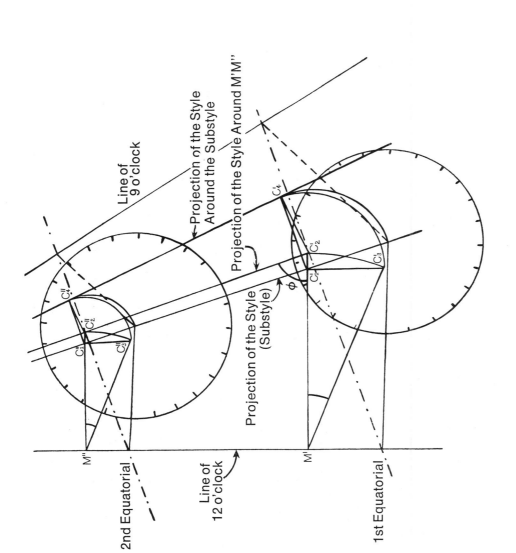

FIGURE 65   Once the auxiliary equatorial dials were in position, only the hour-line of 9 o'clock was drawn. The angles at M' and M'' equal 90—d (complement of angle's declination)

the point c_2' on the extension of $M'c_1'$. This indicates the point of the style of which c_1' is the projection around the noon hour-line. c_2' lies on the projection of the style around this same noon line, making an angle of $90° - \phi$ (the complement of the latitude) with it. So now we can draw it as the straight line $c_2'c_2''$, the point c_2'' being its intersection with the perpendicular to $M'M''$ in M''. An arc of circle of radius $M''c_2''$, with M'' as its centre, gives c_3'' on one side of the angle in M''; we obtain c_1'' by drawing a perpendicular to $M''c_2''$. The straight line $c_1'c_1''$ is the substyle, which is the projection of the style.

We may now repeat in the upper and lower portions the standard steps previously described on p. 61.

E USING MATHEMATICS The formula $\tan \beta = \sin d \cot \phi$ (see p. 62) yields the angle made by the substyle with the noon line. Since we cannot use the centre c, we anticipate the *épure* by a sketch showing the horizontal straight line MS, which has an arbitrary length, in the lower portion of the table, M being a point of the noon line and S a point of the substyle (figure 66). The distance CM is found between the centre c and M by

$$CM = (MS \tan \phi)/\sin d.$$

On the other hand,

$$\tan z = \cos \phi/(\cos d \cot HA + \sin d \sin \phi)$$

gives the angles made by the hour-lines with the noon line and it allows us to obtain the points M_1, M_2, etc. of the intersections of these lines with MS by calculating

$$MM_1 = (CM \cos \phi)/(\cos d \cot HA + \sin d \sin \phi).$$

The angles z_1, z_2, ... being known, we may now trace the hour-lines to the points M_1, M_2, etc.

Once the sketch is completed, the outline of the dial becomes obvious and the final *épure* may be executed by using the results shown above.

Oriental and Occidental Dials

These appear to be declining dials, but the specific value of their declination puts them in a separate class.

Their construction is much easier because the style is parallel to the wall and its placement presents no difficulty since its angle with the horizontal is ϕ. On the other hand, the hour-lines are parallel to the style and to each other since the centre of the dial lies at infinity. Therefore, we need only one point for each of them. In such dials, the six or eighteen hour-line coincides with the substyle.

The previous procedures can now be used to construct them (figure 67).

A USING AN AUXILIARY EQUATORIAL DIAL The 6 or 18 hour-line of the auxiliary dial is directed towards the substyle. The extensions of the other hour-lines indicate the desired points on the wall.

B USING AN AUXILIARY HORIZONTAL DIAL The remarks in A apply.

C USING AN ÉPURE The projection of the equatorial dial around a straight line xx' which is equatorial and perpendicular to the direction of the style is achieved simply by drawing a perpendicular cc' on the point c of the intersection xx' with the substyle, cc' being equal to the height of the style above the plane. cc' is a 6 or 18 hour-line of the equatorial dial, depending on whether the dial being built is oriental or occidental. The extensions of the other hour-lines of the equatorial dial give on xx' the points of the hour-lines to be traced.

D USING MATHEMATICS Here everything becomes simple. $ck_1/h = \tan HA_1$, or $ck_1 = h \tan HA_1$ yields ck_1 and therefore the position of k_1, if h is the height of the style.

The formula indicates that the dimensions of the dial are functions of h, i.e., of the distance of the style from the wall.

General Remarks

It is always necessary to know the earliest and latest hours to be inscribed on the dial, whether it be horizontal, equatorial, or slightly declining.

In general, the hours will not exceed nor be very much behind the hour of sunrise or of sunset on the day of the summer solstice. The hour-angle of the sun for that instant is given by

$$\cos \mathrm{HA} = \tan \phi \tan \mathrm{D},$$

or more specifically by

$$\cos \mathrm{HA} = \tan \phi \tan 23° \ 27',$$

in which ϕ is the latitude. Table 5 gives HA for various values of the latitude using this formula.

In the course of actual construction or in making an *épure*, the last hour to be traced on the side opposite to the declination is the one for which the hour-line of the auxiliary equatorial dial used in the construction becomes parallel with the surface of the table. This observation is valid for inclined dials (see chapter four).

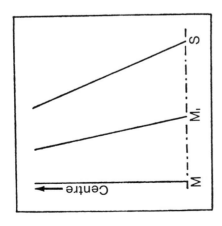

FIGURE 66

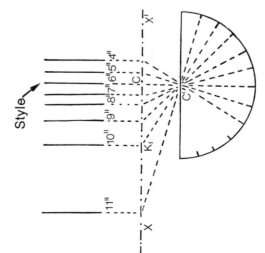

FIGURE 67

CHAPTER FOUR

Inclined Dials

1 GENERAL REMARKS

Inclined dials, sometimes called reclining dials, are not now often made. They seem to belong to the past. We come across them sometimes carved on the faces of stone polyhedra in the gardens of old castles or convents. A famous example found on a terrace of the gardens of the convent of Mont Sainte-Odile in Alsace will be mentioned later (p. 118).

The inclination of the dial seems to make its construction and the calculation pertaining to it more involved. The ancient builders did not encounter any greater difficulty here and seem rather to have been attracted. In any event, we shall see later that the problems posed by these dials are among the most interesting in the study of gnomonics. They therefore deserve a special chapter.

The inclination of the dial, as defined previously (p. 37), modifies the disposition of the hour-lines whose construction is, however, still carried out in accordance with the same principles.

As for the vertical dials, we shall distinguish between the non-declining and the declining dials.

2 MERIDIONAL OR SEPTENTRIONAL INCLINED DIALS

The construction of these dials presents no particular difficulty. Their noon line coincides with the substyle and with the line of greatest slope which is perpendicular to the horizontal and to which the equatorials are parallel.

Figure 68, which shows a portion of the meridian where the dials τ and τ' are inclined towards the north and the south, indicates that the angle between the style and the substyle becomes $\phi - i$ and $\phi + i$ respectively, ϕ being the latitude and i the inclination. The essential task consists then in constructing the dials τ_1 and τ_1' which are horizontal to and identical with those previously described, but have been brought to the respective latitudes of $\phi' = \phi - i$ and $\phi + i$. It has been assumed in the two previous cases that the face of the dial looked skyward. When this is not the case, i.e., when the dial faces the ground, the construction will be identical but the labelling of the hour-lines will be inverted.

We note that $\phi - i$ may become negative; ϕ' corresponds

PLATE 26 Elevation dial (eighteenth century). Its tracing is the unfolding of the one on a shepherd's dial; it does not have a scale for the elevation of the sun. (*Photograph: Musée de la Vie wallonne, Liège*)

PLATE 27 Unfolding of the cylinder of a shepherd's dial according to an engraving found in the book of Bedos de Celle. We note that the curves are the same as those on the elevation dial of Plate 26.

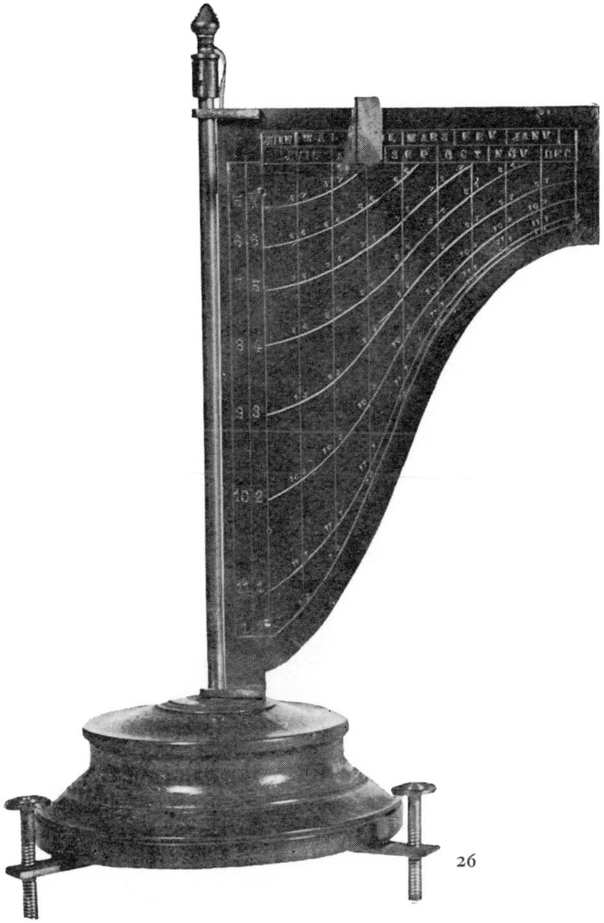

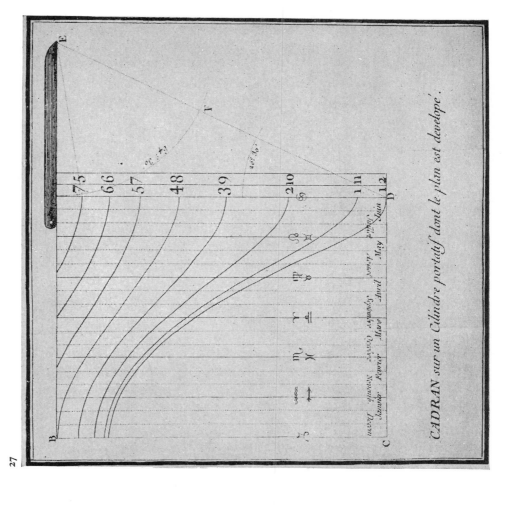

26

27

B
E
75
66
57
48
F
39
210
1 11
Janvier
Fevrier
Mars
Avril
May
Juin
C
1 2
D
CADRAN sur un Cilindre portatif dont le plan est developé.

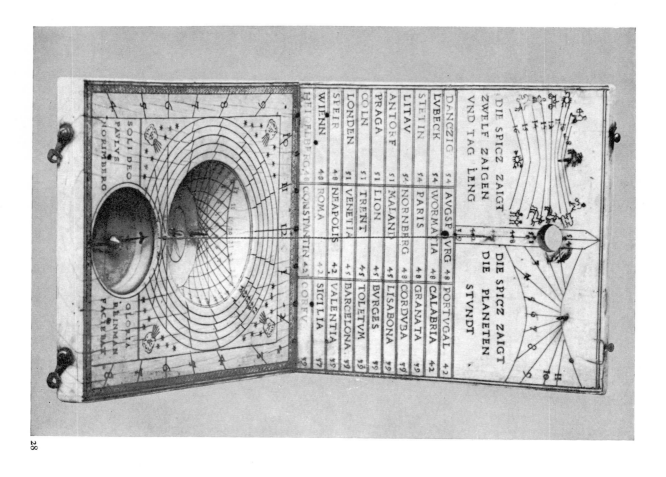

28

29

PLATE 28 Horizontal dial made of ivory in the shape of a diptych originating from Nuremberg; eighteenth century (*Photograph Musée de la Vie wallonne, Liège*)

PLATE 29 Cubic dial with a compass and a plumb-line (eighteenth century). The tables are printed on paper and they are glued to the wood. (*Photograph: Musée de la Vie wallonne, Liège*)

31

32

PLATE 30 Various types of folding
equatorial dials and two horizontal dials.
Only the instrument at the upper left
does not carry a compass, eighteenth
century. (*Photograph: Strasbourg Museum*)

PLATE 31 Pocket equatorial dial folded for
transportation (*Photograph: Dr Kühnelt,
Innsbruck*)

PLATE 32 Pocket astronomical ring folded
for transportation, eighteenth century,
German origin (*Photograph: Dr Kühnelt,
Innsbruck*)

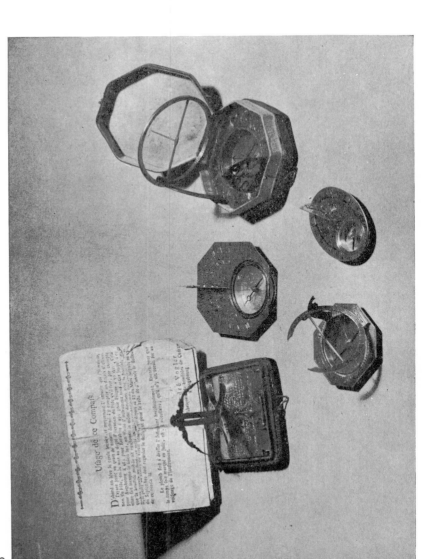

30

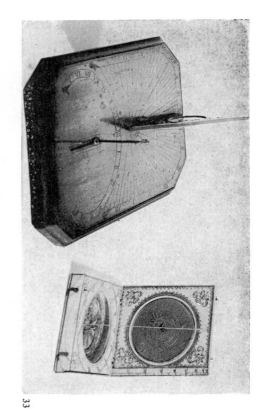

PLATE 33   On the left, a horizontal dial and an analemmatic dial drawn on the same table in order to allow the orientation of the instrument without the help of a compass. On the right, the same combination on a diptych dial made of ivory, but the analemmatic dial located at the bottom of a circular hollow carries a movable ellipse. (*Photograph: Musée de la Vie wallonne, Liège*)

PLATE 34   Acoustical horizontal dial. When the instrument is properly oriented and when the supports of the lens make the proper angle $(90° − \phi + \text{D})$ with the table, the hour of noon is announced by a cannon shot. (*Photograph: Musée de la Vie wallone, Liège*)

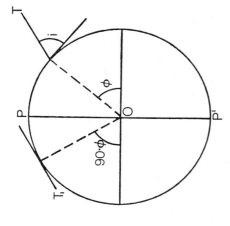

FIGURE 69

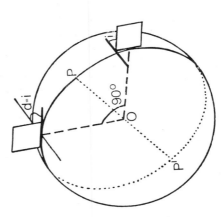

FIGURE 70

then to a southern latitude. Similarly $\phi + i$ may exceed 90°; in this case the new latitude ϕ' equals $180° - (\phi + i)$.

In each case a sketch helps to avoid mistakes (figure 69), the trace of the table T_1 of the horizontal dial being parallel to the table T of the inclined dial and tangent to the circle of the meridian.

3 ORIENTAL AND OCCIDENTAL DIALS

Here again, we may modify the problem by displacing the dial along the meridian by an angle of 90°, keeping the dial parallel to itself. In this way an occidental dial inclined by 35° becomes a vertical dial declining by 35° towards the SW (figure 70) and built for a latitude of $\phi = 90° \pm \phi$. Its construction is, therefore, reduced to one of the more simple cases which we have studied.

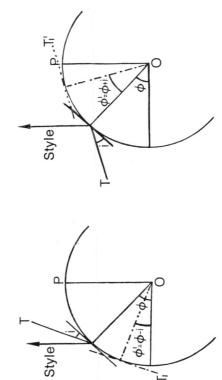

FIGURE 68

4 POLAR DIALS

These dials belong to the particular class of inclined meridional or septentrional inclined dials in which the table is parallel to the axis of the poles, i.e., whose inclination equals the latitude.

The construction is exceedingly simple, reminiscent of the occidental or oriental dials, because here the style is also parallel to the table, but, in addition, the dials are no longer declining and the hour-lines are evidently parallel to each other and symmetrically placed with respect to the noon line. Figure 71 illustrates their construction principle.

BC is the style supported by the rods AB and CD; as usual, we fix on it an equatorial dial whose table is perpendicular to it. The straight line AD is the substyle which happens to be also the noon line. The equatorial dial is rotated around the style till one of its lines falls along AD. The other lines are extended so that we obtain points on the table through which straight lines parallel to AD, the desired hour-lines, are traced.

For this construction it suffices, with the help of an *épure*, to project the auxiliary equatorial dial round the equatorial XY, so that during the projection the centre describes an arc of a circle having the height AB of the style as a radius.

To calculate the distance between the hour-lines and the sub-style, we use the formula

$$x = \text{OP} \tan \text{HA},$$

where HA is the hour-angle of the sun.

Polar dials may be visualized with any azimuth. Their hour-lines remain parallel to the axis of the poles and their construction, with the help of the equatorial dial, is achieved by orienting the noon line of the latter along the axis of the meridian.

5 DECLINING INCLINED DIALS

A. *Installation of the Style*

The problem of the positioning of the style may baffle some readers because of the inclination of the table. In order to avoid mistakes, the plane of the meridian is represented by a plumb-line suspended above the centre of the dial and by a horizontal thread, touching the plumb-line and drawn exactly in the north-south direction.

The axis of the style which should lie here must therefore make an angle of $90° - \phi$ with the plumb-line. On the other hand, the noon line is no longer perpendicular to the horizontal straight lines on the table. Since this is essential, another plumb-line is moved along the meridian line. In this way the desired noon line is obtained. It is wise to check this with the help of a watch set temporarily on the true time. The shadow of the plumb-line suspended above the centre of the dial must coincide with the noon line obtained.

B. *Sketching of the Hour-lines Following the Usual Procedures; Auxiliary, Equatorial or Horizontal Dials*

Once the style is positioned and the noon-line traced, the application of the classical methods presents no particular difficulty (see p. 47).

C. *Drawing the Dial with the Help of an Épure*

The *épure* of an inclined declining dial is really not more difficult than that of a vertical declining dial, in spite of appearances. In-

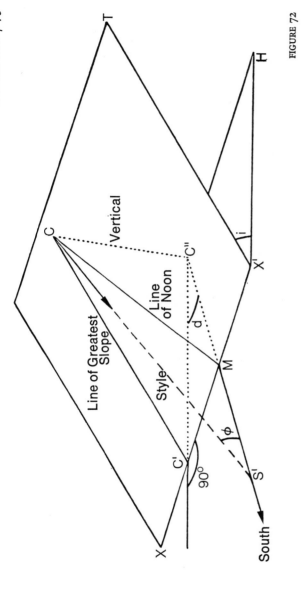

FIGURE 72

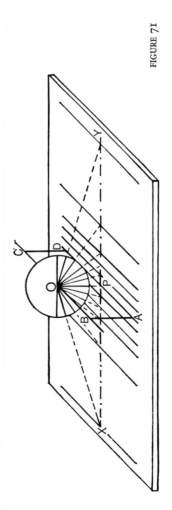

FIGURE 71

deed the same reasoning comes up repeatedly. However, the inclination of the table makes necessary additional operations; in order to keep in sequence the operations to be performed and to help the reader follow them, we shall describe them step by step.

Let us first look at figure 72, where T represents the table of the dial, H the horizontal plane with which T makes the angle of inclination i. Let ϕ be the latitude of the locality and c the point chosen in the plane T which will serve as the centre of the dial to be drawn. From c we draw the perpendicular to H and label its foot c". We also drop the perpendicular cc' to the line xx', marking the intersection of the planes T and H. The plane cc'c" is perpendicular to xx' and cc' is the line of greatest slope on T going through c.

The plane of the meridian going through cc" makes with c"c' an angle equal to the declination d of the style, which is equal to the angle made by xx' with the east-west axis, it is understood that the style cs' lies in this plane and that its axis makes an angle equal to $90° - \phi$ with cc".

Now that we have the data in perspective, we may tackle the épure (figure 73).

First we attempt to construct the point M which determines the plane of the meridian and noon line CM which, we know, is no longer perpendicular to the horizontal xx' as in the vertical dials.

A projection of the triangle cc"c' (figure 72) around the line of greatest slope gives (figure 73) the right-angled triangle cc"c' (in which the angle in c equals $90° - i$). In order to obtain M, we project the right-angled triangle c'c"M (figure 72) on to c'c"M' (figure 73) in which the angle in c" equals the declination d. M' is brought on to xx' by an arc of circle of centre c' and radius c'M'. The noon line CM can now be traced.

In order to find the position of the style which lies in the plane cc"M in figure 72, we project this plane, using CM as a hinge. The

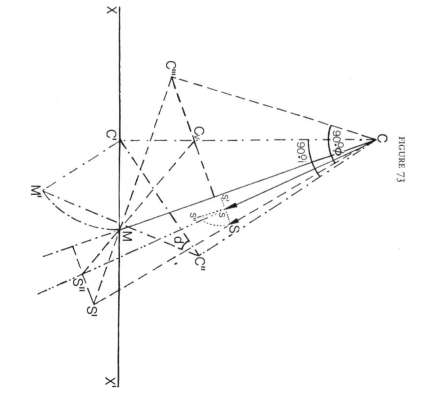

FIGURE 73

triangle obtained, CC'''M, has a right angle in C''' and we know that the length of its side C'''M equals the length C''M' (figure 73) obtained from the previous projection. It is, therefore, easy to construct. If the angle C'''CS is set equal to 90° − φ, the projection of the axis of the style on to CSS' is obtained, S' being the intersection of this axis with the extension of C'''M.

In order to retrace the projection, it suffices to drop perpendiculars to the hinge CM from C'' and S' respectively. The first one gives C$_v$ on CC' and, since the points C'', M, and S' lie on the same straight line (figure 72), we obtain the point S' by extending C$_v$M to the perpendicular to CM issuing from S' (figure 73). CS' is the projection of the axis of the style. If we place the real length CS of the style on the projection CS', we obtain the projection CS with the help of a perpendicular to CM from S.

Now the height of the tip of the style projected in S, above the plane T, must be found. Here again a projection will help us; this time we carry it around SS. Indeed we know the real distance separating the tip S of the style from the noon line CM in the projection CMS': it equals S'S if we label the intersection of SS with CM by S'.

If we now raise a perpendicular to SS at S, an arc of circle of radius S'S and centre S' gives the point S''. Then SS'' is the desired elevation.

From now on, the construction of the *épure* may proceed as for the previous dials, i.e., by the usual projection of the vertical plane containing the style and the substyle, around the latter, i.e., around the straight line CSS''' in figure 73.

The perpendicular SS''' is raised at S (figure 74) equal to the height of the tip of the style above T and CS''' becomes the real length of the style, thus indicating another method of obtaining the point S'''. The perpendicular on to CS''' in S''' gives the point E on CS which is a point of the equatorial and this latter is the

FIGURE 74 For the sake of clarity in the figure, only the hour-lines between 11 A.M. and 2 P.M. have been drawn

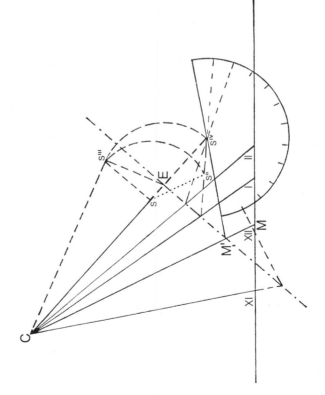

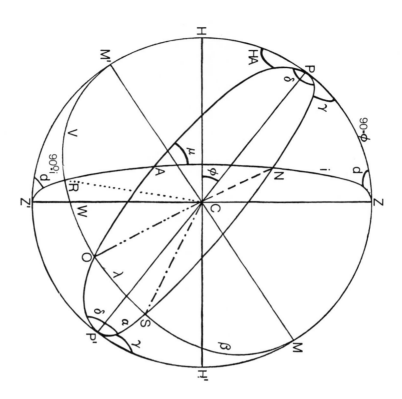

FIGURE 75

perpendicular to cs in E. As usual, we assume that the auxiliary equatorial dial is applied at the tip s''' of the style and we project it around the equatorial. An arc of a circle of centre E and radius Es''' gives the centre of the dial's projection in s^{iv}. By joining this point to M', the intersection of the equatorial with CM, the straight line $s^{iv}M'$ becomes the projection of the noon line of the equatorial dial. It suffices now to construct the other hour-lines of the latter and to extend them to the equatorial in order to obtain the corresponding points of the dial to be built. We complete the drawing by joining them to C.

D *Declining and Inclined Dials by Mathematics*

The Problem

In the simple case of the horizontal dial we have shown how the elements of a dial can be obtained by using mathematics and applying elementary reasoning. This type of calculation is usually given in current literature for all the particular cases, especially for horizontal and vertical dials, whether declining or not.

If we approach the problem of the calculation of the elements of dials that are simultaneously inclined and declining, the simple procedure followed in the first example becomes intricate and leads to formulae that are too complicated for simple interpretation. The best and most rational approach to the solution of the problem is to go back to basic principles and to tackle it within the system of reference of the celestial sphere. This makes the problem clearer and more general and the solution to which it leads applies simultaneously to all the classical dials. This is why we have not treated the mathematics of each of the individual dials described, as we did for the first one (as an example). Since then such manipulations have been avoided and we have restricted ourselves to deducing from the general formulae those relevant to the case being treated. These general formulae will now be given in the following pages.

Although the solution belongs to the field of elementary mathematics, it is not obvious at first view. The principal reason is that it leads to general formulae, giving the elements of any type of dial. The second reason is the hope that it will be welcomed by those who enjoy solving the interesting problems raised by gnomonics and that it may lead to further research.

The Calculation of the Angles between the Hour-lines and the Noon Line

Let PZP′z′ be the celestial sphere (figure 75), C being its centre and also the centre of a declining and inclined dial; the great circle of diameter M′M is the plane of the dial. HH′ is the plane of the horizon, PP′ the line of the poles and also the style of the dial, and ZZ′ the vertical at the locality. If CN is the normal to the plane of the table M′M, then the angle ZN equals the inclination i of the dial. If the vertical plane ZNZ′ goes through CN, the angle PZN is the declination of the dial. If the shadow of the style falls on a point O of the dial, POP′ is the hour-circle of the sun and the angle MPO is the hour-angle HA of the latter. Let MCM′ be the noon line of the dial.

The shadow O makes an angle OCM = z with the noon line. This angle OCM with the noon line is determined by the arc MO or MO′, the elevation of which is the goal of all our calculations. The great circle ZNZ′ divides it into two parts v and w, giving $z = v + w$. The intersection R of the plane of the dial with the vertical plane ZNZ′ is a right angle since the latter contains the vertical CN to the plane M′M of the table. An examination of the figure shows that v is the angle between the line of noon and the line CR of the plane M′M which is its line of greatest slope, and w is the angle between the hour-line corresponding to the hour-angle HA of the sun and this same line of greatest slope.

Let A be the intersection of the vertical ZNZ′ with the hour-circle POP′. In the right-angled triangle M′RZ′, we have

$$\tan v = \tan d \sin \text{RZ}'.$$

Since RZ′ = $90° - i$, this becomes

$$\tan v = \tan d \cos i.$$

The formula confirms the obvious fact that v is independent of time. In the triangle ARO, which has a right angle in O, we have

$$\tan w = \sin \text{AR} \tan \mu,$$

if we call μ the angle PAZ. We note that, in the triangle PZA,

$$\cos \mu = \cos d \cos \text{HA} + \sin d \sin \text{HA} \sin \phi, \qquad (1)$$

where ϕ stands for the latitude PCH, the arc PZ then being equal to $90° - \phi$. Also

$$\text{ZA} + \text{AR} = 90° + i,$$

$$\text{AR} = 90° + i - \text{ZA},$$

and therefore $\tan w = \tan \mu \cos (i - \text{ZA})$. We can thus write

$$\tan w = \frac{\cos i \cos \text{ZA} \sin \mu + \sin i \sin \text{ZA} \sin \mu}{\cos \mu} \qquad (2)$$

But in triangle APZ, we have

$$\frac{\sin \mu}{\cos \phi} = \frac{\sin \text{HA}}{\sin \text{AZ}}$$

i.e.

$$\sin \text{AZ} \sin \mu = \cos \phi \sin \text{HA}. \qquad (3)$$

Multiplying this equation by cot AZ, we obtain

$$\cos \text{AZ} \sin \mu = \cos \phi \sin \text{HA} \cot \text{AZ}. \qquad (4)$$

If we replace $\cos \text{ZA} \sin \mu$ and $\sin \text{ZA} \sin \mu$ in (2) by their values in (3) and (4), we have

$$\tan w = \frac{\cos i \cos \phi \sin \text{HA} \cot \text{AZ} + \sin i \cos \phi \sin \text{HA}}{\cos \mu} \qquad (5)$$

To obtain cot AZ, we apply the cosine formula to the triangle PZA:

$$\cos d \sin \phi = \cot \text{ZA} \cos \phi + \sin d \cot \text{HA}$$

which yields

$$\cot \text{ZA} = \cos d \tan \phi - \frac{\sin d \cot \text{HA}}{\cos \phi}.$$

Substituting this value into (5), after some simplifications, we obtain

$$\tan w = \frac{\cos i \sin \phi \cos d - \cos i \cot \text{HA} \sin d + \sin i \cos \phi}{\cos u / \sin \text{HA}}.$$

We again use (1), which gives

$$\cos u / \sin \text{HA} = \cos d \cot \text{HA} + \sin d \sin \phi,$$

so that

$$\tan w = \frac{\cos i \cos d \sin \phi + \sin i \cos \phi - \cos i \sin d \cot \text{HA}}{\cos d \cot \text{HA} + \sin d \sin \phi}.$$

We now have explicit values for tan v and tan w, from which we can infer v and w as well as $z = v + w$.

Despite appearances, this calculation is simple since v is calculated once and for all, while for w there are only two factors containing the variable term HA.

Calculation of the Angle between the Noon Line and the Line of Greatest Slope

We recall the first formula obtained above:

$$\tan v = \tan d \cos i,$$

which gives the angle v between the line of greatest slope CR and the line of noon where d and i are arbitrary. This formula is of fundamental importance for the calculation of the elements of a declining and inclined dial, because it determines the position of the noon line which serves as a reference co-ordinate for the sketching of the other hour-lines.

Calculation of the Angle between the Style and the Substyle

In order to avoid confusion, we recall that the axis PP' is not perpendicular to the plane MM' in our drawing; the plane MM' should not be confused with the plane of the equator which is not shown.

The plane perpendicular to the plane of the dial MM', containing PP' and CN, the normal to that plane, cuts it along CS, the substyle.

CS makes an angle with the noon line which is measured by the arc SM and which we call β. The style PP' makes an angle α with the substyle, which has the arc P's for measure. NS, by definition, is a right angle.

PN is the complement of SP' = α. In the triangle PNZ, we have

$$\cos \text{PN} = \cos (90° - \phi) \cos i + \sin (90° - \phi) \sin i \cos d; \qquad (6)$$

in other words,

$$\sin \alpha = \sin \phi \cos i + \cos \phi \sin i \cos d.$$

Calculation of the Angle between the Substyle and the Noon Line

Let us look again at the triangle ZPN. We note that

$$\cot \angle \text{ZPN} \sin d = \cos \phi \cot i - \sin \phi \cos d$$

or

$$\cot \angle \text{ZPN} = \frac{\cos \phi \cot i - \sin \phi \cos d}{\sin d}. \qquad (7)$$

$$\tan \angle \text{ZPN} = \tan \angle \text{SP'M} = \frac{\sin d}{\cos \phi \cot i - \sin \phi \cos d}.$$

This may be written as

$$\cot \angle \text{ZPN} = \frac{\cos \phi \cot i - \sin \phi \cos d}{\sin d}.$$

If, as before, we call S'M, in the right-angled triangle SP'M we have

$$\frac{\tan \beta}{\sin \alpha} = \tan \text{SP'M, or}$$

$$\frac{\tan \beta}{\sin \alpha} = \frac{\sin d}{\cos \phi \cot i - \sin \phi \cos d}.$$

For sin α we take the value found in (7) and, therefore,

$$\tan \beta = \sin i \sin d \frac{\tan \phi \cos i + \sin i \cos d}{\cos i - \tan \phi \cos d \sin i}.$$

Calculation of the Hour-angle of the Sun at the Instant when its Shadow falls on the Substyle

The shadow falls on the substyle at the instant when the hour-angle of the sun equals MPS. We call this angle γ. We then have

$$\tan \gamma = \tan \beta / \sin \alpha$$

or, after some simplifications,

$$\tan \gamma = \frac{\sin i \sin d}{\cos i \cos \phi - \sin \phi \cos d \sin i}.$$

Arrangement of the Hour-lines with Respect to the Substyle

If we reckon time from the substyle (which would be highly impractical), the angle of the desired hour-line would be OPS $= \delta$. The shadow would make an angle OCS $= \lambda$ with the substyle, and in the triangle OSP (which has a right angle in S)

$$\tan \lambda = \sin \alpha \tan \delta.$$

Since α is independent of time, the formula expresses the fact that the hour-lines are arranged symmetrically with respect to the substyle.

Particularization to some Special Cases

The corresponding formulae pertaining to the particular cases are deduced by assigning to d or i their appropriate values. These are shown in the following table.

SPECIFIC VALUES OF i, d, α, β, AND γ FOR SOME PARTICULAR TYPES OF DIALS

Name	Inclination (i)	Declination (d)	Angle between the shadow and the noon line (z)	Angle between the style and the substyle (α)	Angle between the noon line and the substyle (β)	Hourly angle of the sun at the instant when the shadow falls on the substyle (γ)
NON-DECLINING DIALS						
Horizontal	$0°$	$0°$	$\tan z = \sin \phi \tan \text{HA}$	ϕ	$0°$	$0°$
Meridional vertical	$90°$	$0°$	$\tan z = \cos \phi \tan \text{HA}$	$90° - \phi$	$0°$	$0°$
Septentrional vertical	$-90°$	$0°$	$\tan z = \cos \phi \tan \text{HA}$	$90° + \phi$	$0°$	$0°$
Equatorial	$90° - \phi$	$0°$	$z = \text{HA}$	$90°$	$0°$	$0°$
Polar	$-\phi$	$0°$	$z = 0$	$0°$	$0°$	$0°$
Meridional inclined	Arbitrary	$0°$	$\tan z = \sin (\phi + i) \tan \text{HA}$	$\phi + i$	$0°$	$0°$
DECLINING DIALS						
Vertical declining	$90°$	Arbitrary	$\tan z = \dfrac{\cos \phi}{\cos d \cot \text{HA} + \sin d \sin \phi}$	$\sin \alpha = \cos d \cos \phi$	$\tan \beta = \sin d \cot \phi$	$\tan \gamma = \tan d \sin / \phi$
Oriental	$90°$	$90°$	$\tan z = \cot \phi$	$0°$	$\tan \beta = \cot \phi$	$90°$
Occidental	$90°$	$-90°$	$\tan z = \cot \phi$	$0°$	$\tan \beta = \cot \phi$	$90°$

ϕ is the latitude and v, the angle between the noon line and the line of greatest slope, is zero for all the dials mentioned except the occidental and oriental dials.

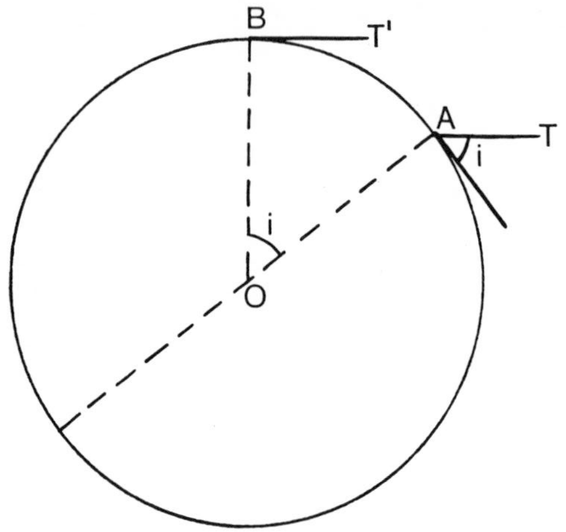

Reduction of the Construction of a Declining and Inclined Dial to that of a Horizontal Dial

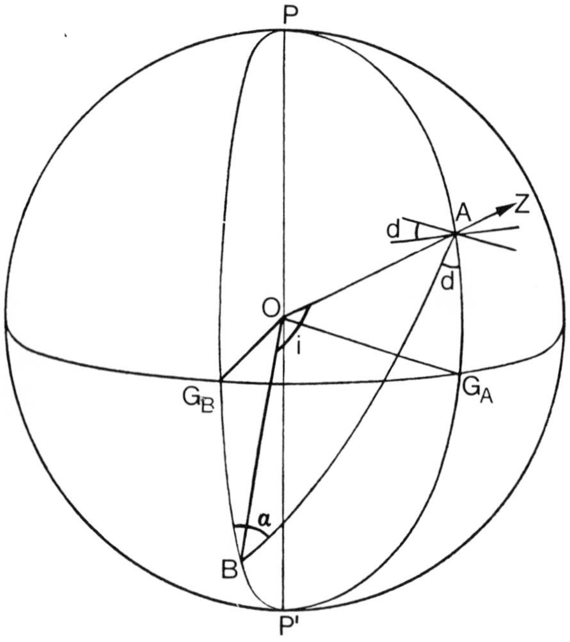

We have said previously (p. 70) that the construction of a meridional or septentrional dial having an inclination i at a latitude ϕ could be reduced to that of a horizontal dial visualized at a latitude which, in principle, should be equal to $\phi - i$ or $\phi + i$ depending on the case considered.

FIGURE 76

There is nothing to stop us from studying the solution to the problem of an inclined and declining dial from an analogous point of view. Consider a section of the earth's sphere as the plane of a great circle containing the normal to the plane of the table of the dial. This plane cuts the table of the dial along the line of greatest slope (figure 76). Now imagine that the table T is displaced parallel to itself by an angle i from A to B along the great circle; it occupies the position T' in B, that of a horizontal dial (the displacement being performed in the direction opposite to the inclination i). If ϕ_A and g_A are the co-ordinates of the point A, the latitude is obtained in the spherical triangle PAB (figure 77), by using the formula

$$\cos PB = \cos PA \cos AB + \sin PA \sin AB \cos \angle PAB.$$

If we call the declination of our dial d and if we substitute the angular measure of the arcs in the triangle, we obtain

$$\sin \phi_B = \sin \phi_A \cos i + \cos \phi_A \sin i \cos d,$$

ϕ_B being the latitude of the point B.

The difference in longitude between A and B is given by

$$\sin PA \cot AB - \sin \angle PAB \cot \angle BPA = \cos PA \cos \angle PAB,$$

or

$$(\cos PA \cot i + \sin \phi_A \cos d = \sin d \cos(g_B - g_A)$$

where g_B is the longitude of the point B, and

$$\tan(g_B - g_A) = \frac{\sin d}{\cos \phi_A \cot i + \sin \phi_A \cos d}.$$

FIGURE 77

The plane of the great circle AB goes through A in the line of greatest slope of the dial. The noon line of the dial, located in the meridian, coincides with the substyle in B. The angle between these lines, PBA or α, is determined by using the formula

$$\sin \text{AB} \cot \text{PA} - \sin \angle \text{PAB} \cot \angle \text{PBA} = \cos \text{AB} \cos \angle \text{PAB}$$

or

$$\sin i \cot (90° - \phi_\text{A}) - \sin (90° - \phi_\text{A}) \cot \alpha = \cos i \cos (180° - d),$$

i.e.,

$$\tan \alpha = \cos \phi_\text{A} / (\sin i \tan \phi_\text{A} + \cos i \cos d).$$

We have now found the angle between the substyle and the line of greatest slope. When it is noon on the dial B, the shadow falls on the substyle. When it is noon at A, the shadow of the style makes an angle ϕ_B with this substyle.

Using these results, the construction of the dial is reduced to the following operations.

(1) The sketching of the substyle which makes an angle α with the line of greatest slope; the installation of the style which makes an angle ϕ_B with the substyle and the plane of the table.

(2) The sketching of an angle which has the centre of the dial for a summit and which makes an angle $(g_\text{B} - g_\text{A})$ with the substyle. The angle must have the proper orientation, depending on whether the difference in longitude is east or west, i.e., on whether the declination of A is west or east. The open side of this angle is the noon line of the inclined dial.

(3) The setting of the auxiliary equatorial dial, its noon line being rotated till it meets the noon line just traced, and the sketching of an equatorial and of the hour-lines of the dial according to the usual procedure.

CHAPTER FIVE

Solar Calendars

I VARIATION OF THE SUN'S DECLINATION AND THE WAY IT AFFECTS A SUNDIAL

We have seen that, because of the earth's rotation, the sun seems daily to describe a circle around the axis of the poles and that, because of the annual trajectory of the earth around the sun (during which the direction of the axis of rotation of the earth is practically fixed), the declination of the sun varies between $+23°\ 27'$ and $-23°\ 27'$. If the earth is labelled T and the sun s in figure 78, the surface generated daily by the straight line TS is a cone of revolution with a large angle of aperture, whose shape varies between the two extremes ETE' and HTH' which it reaches during the summer and winter solstices; it goes through some intermediate forms and during the equinox is reduced to a plane. During the daily pass of the sun, the shadow of a point on which it shines (tip of the style) describes a sort of curve on the table. It is the intersection of the surface of the daily cone with that of the dial.

We know that the intersection of a plane with a cone of revolution is a circle, an ellipse, a parabola, or a hyperbola, depending on the angle which it makes with the axis of the cone (figure 79).

If D is the declination of the sun and α the inclination of the plane of the table of the dial, the curve described by the shadow of

the style's tip on the table is a hyperbola if D is less than α, a parabola if D equals α, and an ellipse if D is greater than α.

In the polar zones, arctic or antarctic, for a horizontal dial, the shadow of the tip of the style describes ellipses that become circles if the dial is exactly on the pole; for a vertical dial, we obtain hyperbolas. Inversely, in the torrid zone (between the tropics), a horizontal dial shows hyperbolas while a vertical dial may show circles, ellipses, parabolas, or hyperbolas, depending on its declination with the first vertical. In the temperate zones, the curves obtained are hyperbolas on both vertical and horizontal dials. Various declinations give various results, the four types of curves being always possible. For an equatorial dial which is the equivalent of a horizontal dial situated at one of the poles, the curves are concentric circles.

Theoretically the daily displacement of the curve of the shadow on the dial supplies a means for determining the date. To effect this, the curve should be traced for each day, but this is not practical. Nevertheless, in the past, certain noteworthy dates — religious holidays for instance, and especially the dates of the entry of the sun into the various signs of the zodiac — were indicated in this way on many dials. In practice, this involves the tracing of the daily curves for the twenty-first day of each month. Since these curves overlap in pairs, corresponding to equal solar declinations, only seven curves in all are needed.

Indeed the curves for February 21 and October 21, January 21 and November 21, etc. coincide. The curves for March 21 and September 21, namely the equinoxes, during which the declination of the sun is zero, mark the limits of the change in the convexity of the hyperbolas: these curves become overlapping straight lines. They are represented on a dial by the equatorial which we have mentioned many times in the preceding chapters.

The curves so defined are called *monthly* or *daily arcs* or *curves* or *arcs* or *curves of declination*. They determine the declination of the sun as well as the date, since there is a relation between them.

These possibilities have been turned to advantage in order to render sundials more interesting as well as more instructive, since they allow us to follow the movement of the earth around the sun. They have been embellished by the addition, beside each curve, of the sign of the zodiac which the sun is about to enter, and this facilitates the reading of the date (figure 80).

Table 4 gives the mean value of the declination of the sun during the course of the year. Its use in the tracing of the daily arcs guarantees the necessary accuracy for gnomonics.

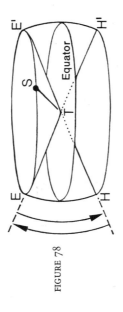

FIGURE 78

2 TRACING OF THE CURVES OF DECLINATION

The number of cones, whose intersections with the plane of the table are to be determined, number only six, since the central cone, as we have seen, is a cone with a 90° aperture and, therefore, a plane perpendicular to the style and intersecting the table of the dial in the equatorial. The top of these cones, on the other hand, is chosen to be the tip of the style or, sometimes, a little ball stuck on the style whose shadow will supply the desired information.

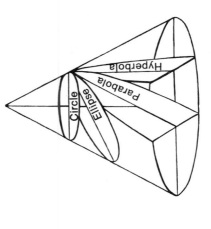

FIGURE 79

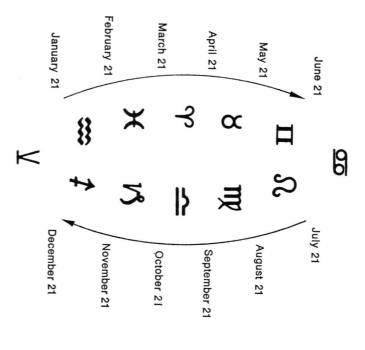

May 21
June 21
April 21
March 21
February 21
January 21

July 21
August 21
September 21
October 21
November 21
December 21

FIGURE 80   The ascent and descent of the sun between the solstices. When the symbols are located between the daily arcs of a dial, their arrangement is symmetric with respect to the tip of the style. It is, therefore, obtained by holding the book upside down (see also Figure 21)

The procedure for making a practical drawing of these curves is extremely simple.

In figure 81, let cs be the style and s its tip, and let us assume that the thin rods sp, sq, and sr represent the direction of the sun's rays at the true noon (the reasoning is the same for any hour) during the time of the summer solstice, the equinox, and the winter solstice, respectively. The extensions of these rays meet the table in p′, q′, and r′. During the sun's daily pass, each of these rays describes a cone of revolution in space by rotating around cs (cs is the axis of the cone while s is its tip); in the course of a day we may, as always, neglect the variation of the declination of the sun and consider it as fixed. In this case, the points p′, q′, and r′ each describe a branch of a hyperbola. At the equinox, the declination of the sun, as we have said previously, is zero, which means that sq is perpendicular to cs and that the point q′ moves along a straight line, namely the equatorial.

On the other hand the angles psq and qsr are known; they are equal to the declination of the sun on the days of the equinox, namely +23° 27′ and −23° 27′. We can then picture a mobile instrument around the style, carrying the rays sp, sq, and sr as well as all the others whose daily hyperbolas should be traced on the dial; as we have already stated, their number is usually reduced to seven, including the straight-lined equinoctial curve.

This instrument, called a *trigon*, has taken an infinite number of forms in the history of gnomonics. Figure 82 shows one of its simple forms in wood. It is fixed to the style by two brass straps in which the style is completely embedded; its extremity touches an iron point t to which a thread is attached. A hollow must be made in the wood of the trigon for the insertion of the style so that its axis is level with the plane of the trigon which carries the scale of angles or the signs of the ecliptic. A fixing screw v, on the underside of the trigon, sets the instrument in position.

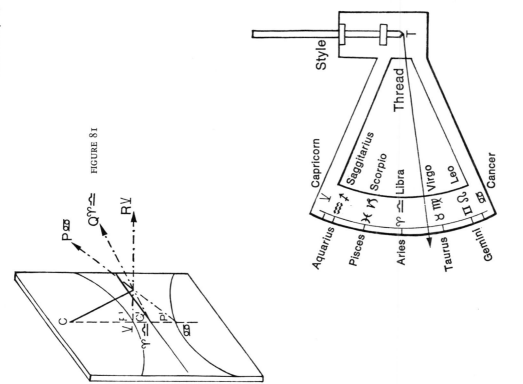

FIGURE 81

FIGURE 82   Wooden trigon

We may then rotate the trigon in as many positions around the style as seem necessary. For each of these positions we obtain a point of the different daily curves simply by stretching the thread flat on its plane along the corresponding graduations. The place where the thread touches the plane of the dial always makes a point of one of the curves.

An ordinary protractor may be transformed into a trigon by a slight modification. It will then supply the curves for any declination of the sun.

When the dial must be drawn with the help of an *épure*, the same results may be obtained by a purely graphic method. As usual, we proceed by projections. Let us assume that we have a dial already built (figure 83) where CS is the projection of the style (the height *h* of its tip above the plane of the table represented by SS' in the projection is already known) and let CH be any hour-line on which we seek the intersection of the daily curves.

Seven lines are to be drawn on the plane defined by the point S and the straight line CH; one of these, the one corresponding to the equinoxes, is perpendicular to the axis CS, while the others make angles with it equal to the respective declinations of the sun on the twenty-first day of the various months.

We drop the perpendicular SK to CH and draw SS′ = *h* perpendicular to SK. Then KS′ is the length in three dimensions of the straight line KS. In order to rotate our plane around CH and obtain the projection S″ of S, we need only draw the arc of the circle S′S″, with centre K, which meets the extension of KS in S″. The projection S″ of the style is therefore CS″. We trace the perpendicular S″D of the equinox in S″, and draw from S″D the angles of the declination whose sides cut CH in the points A, B, O, E, F, and G, which are points of the desired curves. Repeating this construction for the various hour-lines on the dial (and where need be, for extra lines in case some intervals are too far apart), we obtain other sets of

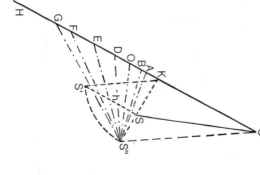

FIGURE 83

points, each of which corresponds to a given curve. The tracing of the curve may be done by hand or more easily with the help of an appropriate tool for drawing curves.

3 THE MERIDIANS

We have seen that the mean time does not coincide with the true time because of the elliptic shape of the earth's orbit and of its inclination to the equator; the difference between these two times is a variable quantity called the equation of time. To obtain the mean local time from the reading of a dial, we must, as we already know, add or subtract this variable quantity from the true time of the dial, depending on whether it is positive or negative. A plot (figure 22) or a table (Table 3) may be used to obtain its value on the day of observation. But it is clear, as we have just seen, that a fairly accurate idea of the date may be gained by looking at the shadow of the style on a dial bearing the daily arcs. The data are inserted on the dial, an operation that makes looking up the equation of time superfluous. The mean time can now be read, in a manner of speaking, directly from the dial.

Indeed to find out when it is mean noon on a given day, it is only necessary to draw a line on the appropriate daily arc from the hour-line of noon and in the proper direction, so that the shadow indicates mean noon at the instant when it reaches the end of the line just drawn. The distance between this line and the noon hour-line is to be for every day, i.e., on every daily arc, equal to the time equation of that day.

Assume that we have drawn on the dial the daily arcs for every day of the year. Using, as we shall explain later, supplementary hour-lines on either side of the noon line, and having drawn the equation of time, we obtain a way of determining the mean local noon from the shadow of the style.

This curve assumes the shape of an 8 as shown in figure 84 to which the name *meridian of the mean time* has been given. Theoretically, in order to have the mean time on the actual hours, we should carry such curves on each hour-line (figure 85). This has, in fact, been done for some rare dials. In practice, we content ourselves with a single graph, from which we need only read off the data to deduce the mean time throughout the year.

In order to obtain the curve, the hour-lines are drawn in pencil at intervals of 5 minutes around the noon line; this gives six lines in all since the equation of time seldom exceeds 16 minutes in either direction. Daily arcs for dates 10 days apart, i.e., two interpolations for each month, may be inserted at sight between the seven main daily arcs without any appreciable error; afterwards the values of the equation of time supplied by the table are transferred to this outline of lines and arcs.

The curve obtained is not symmetrical. It depends on the scale of the hours, i.e., on the width of the hour-angles at the various dates; in a general way its form is more or less modified according to the variable direction of the daily arcs on the surface of the dial. It is not symmetrical even on meridional or horizontal dials, which themselves are symmetrical. The four points of its intersection with the noon line indicate the points and the dates for which the equation of time is zero.

Since each daily arc cuts it at two points, the branches referring to the date in question must be indicated without ambiguity. It is here that the decorative character of the signs of the zodiac have been used in the construction of sundials: they are drawn along the curves at suitable places, preferably near the meridian, at equal distances from the daily arcs which delineate the time of the solar transit into each of them. Sometimes the numbers of the months in Roman numerals or even the names of the months or the seasons are given instead.

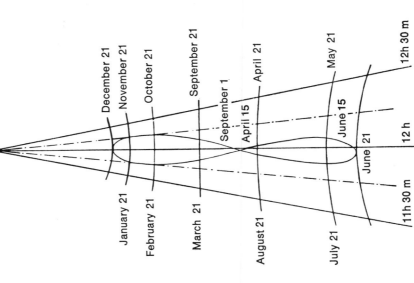

FIGURE 84   The meridian of the mean time

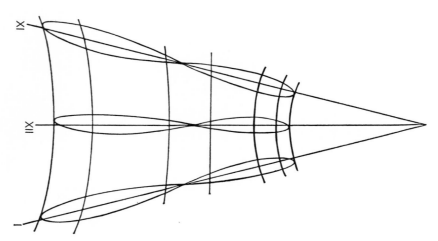

FIGURE 85   Part of a dial in which each hour-line carries a meridian of the mean time (see also Plate 19)

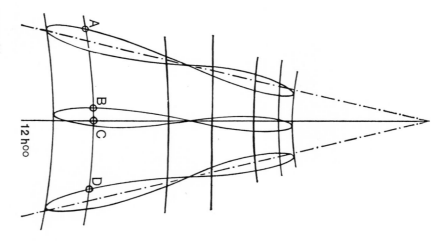

FIGURE 86

Some installations, being noon mark dials of a similar use and looked upon more or less as monuments, were at one time highly regarded for the setting of the first mechanical watches. They will be mentioned later.

The meridian may be used to determine the mean legal time, which differs from the mean local time by the difference in longitude between the local meridian and the meridian of the time zone (on which the watches of the time zone or of the country are set) expressed in units of time. In this case, the curve of the equation of time should be arranged around an hour-line on the dial removed from the noon line by an amount of time equal to this difference in longitude in the appropriate direction (figure 86).

As an example, we take a city in England with an east longitude. We first chose as its standard legal time the winter time based on the Greenwich meridian, then the time used in summer, which as we know is one hour ahead. Suppose that Ramsgate is the town chosen. Its longitude is 1° 25′ E or 5 minutes 40 seconds in time. When it is true local noon, the longitude of 5 minutes 40 seconds must first be subtracted from noon to obtain the true Greenwich time (standard or winter time); the hour-line of 11 hours 54 minutes 20 seconds is therefore the true Greenwich noon.

If, during the construction of the dial, we rotate the auxiliary equatorial dial by 5 minutes 40 seconds or 1° 25′ towards the left, we obtain in the *épure* the proper noon line, so that the passage of the shadow indicates the hour of noon in legal time. It is understood that the auxiliary equatorial dial installed in this way gives as well the other hour-lines with a similar time shift. To use this dial for reading summer time, the numbering of the hours must be increased by one and the meridian of the equation of time carried to the new noon line.

Figure 86 shows side by side, with their meridians, the noon lines of a dial for the legal time, the true local time, and the sum-

mer time, respectively; we note that only the noon line of the true time remains vertical. It is clear that all this information will clutter the dial considerably and may lead to confusion. Only one of these lines is normally retained and, in general, dials indicating the legal time have remained extremely scarce. However, for an area like Finistère in France, where the difference between the true time and the legal time in use – the summer time – is large, their use seems to have some obvious advantage.

4 BABYLONIC AND ITALIC HOURS

The people of Babylon, to whom we owe the division of the day into 24 hours, used to count the hours starting from sunrise. In contrast, the Italians of the Middle Ages started a new day at sunset. Both of these methods of measurement have been in use for a very long time; the Babylonian method was used in Greece and in the Balearic Islands, while the Italic method was still current in Goethe's time; he refers to it in his *Italienische Reise*. All over Europe these methods have been in use at various epochs, and in some countries they were abolished only a little over a century ago, so that countless old European dials are to be found even now, which refer to Babylonic or Italic hours.

In Germany, mainly in Nuremberg and Swabia at first, the custom arose at the time of the Renaissance of counting time in Babylonic hours during the day and Italic hours during the night; consequently the original division into 24 hours was gradually split into two subdivisions of 12 hours each, a custom that has prevailed ever since and which has resisted all attempts at reform. We still find many dials bearing both Italic and Babylonic hours on many cathedrals, in particular those of Strasbourg and Basel. Passers-by may give them a casual glance but are ready to admit that they do not understand the significance of these lines.

It is true indeed that these dials are no longer practical for indicating the time. However, let us picture a dial on which the Babylonic hours are shown with their origin at sunrise and then the Italic hours not labelled in their traditional sequence but in the opposite way, so that the line indicating sunset is the line 0, the previous hour 1, and so on until the hours 13, 14, and 15 are reached on the left-hand side of the dial.

Now by studying the dial with the new lines, we shall obtain for each date indicated by the shadow of the tip of the style:

(1) the number of hours elapsed since sunrise. This is obtained from the Babylonic hours. The first of these, which may not be found on some dials starting from a given value of their declination if it is southwest, is the hour 0, i.e., the Babylonic hour of the sunrise;

(2) the number of hours left till sunset, i.e., the number indicated by the position – in reverse order as we have already said – of the Italic hours;

(3) the length of the day for the date in question obtained by adding the two together. This can be done even if there is no sun and for any day of the year by taking an arbitrary point on the daily arc for which we add up the values indicated by the two types of lines;

(4) the intersections by pairs of the two types of lines situated on the ordinary hour-lines of the dial – the so-called *astronomical lines*.

Other intersections are found between them on the half-hour-lines (usually not drawn) which they make superfluous. Independent of the sun, the dial has, therefore, been effectively transformed into an abacus which may be used to determine the length of the day for any day of the year. In addition, starting from the noon line – on which the two types of lines intersect in pairs – we may determine immediately the true hour of either sunrise or

sunset. There is no doubt that a dial carrying the Babylonic and Italic hours offers a wealth of information which may not be suspected at first sight.

The declination of the sun is practically constant along a given daily arc. The intersections by pairs of the two types of lines must therefore be found automatically on the same daily arc – even if the latter is not indicated on the dial. These intersections show points of additional daily arcs and it is generally understood that the equatorial will carry the intersections of all the lines whose sum of hour value, by pairs, is 12, i.e., the effective length of an equinox day.

5 TRACING OF THE BABYLONIC AND ITALIC HOUR-LINES

As we shall see later, the construction of these lines, although easy in principle, requires much careful work: we must see that the errors arising from translating the measurements of the *épure* to the dial do not accumulate and eventually invalidate the final result. We shall find that the lines are straight and that we need only two points for the drawing of each one.

A. *Practical Tracing*

The practical procedure for drawing the lines with sufficient accuracy is described as follows.

We assume that our dial already carries the ordinary hour-lines, i.e., the so-called astronomical lines, as well as the daily arcs of the two days of the winter and summer solstices, traced with the greatest care. It may be further assumed that we know the time of the sunrise on each of these days. To marshall our ideas, let this time be 7h 55m (true time) for the winter solstice and 3h 52m for the summer solstice. The points corresponding to the hour-lines of 7h 55m, 8h 55m, 9h 55m, etc., on the daily arc of the winter solstice are entered and carefully extrapolated. If greater precision is needed, the corresponding hour-lines and the points of their intersection with the daily arc may be constructed by following the procedures already described. Similarly we draw on the daily arc of the summer solstice the hour-lines of 3h 52m, 4h 52m, 5h 52m, etc. The Babylonic hours desired are the straight lines joining the points of the arc of the winter solstice for 7h 52m, 8h 52m, 9h 52m, etc., respectively, to those of the summer solstice of 3h 52m, 4h 52m, 5h 52m, etc. We obtain the Babylonic hour-lines in this way or, in other words, the lines that indicate the number of hours elapsed since sunrise.

In order to obtain the Italic hour-lines which, of course, indicate the Italic hours or the number of hours which have yet to elapse till sunset, we proceed in an analogous fashion. Indeed, going back to our example, the sun sets at 8h 08m in the evening on the day of the summer solstice and at 4h 05m in the afternoon of the winter solstice. As before, we draw the points of intersection of the hour-lines of 8h 08m, 7h 08m, 6h 08m, etc. to those of 4h 05m, 3h 05m, 2h 05m, etc. on the daily arcs of the summer and winter solstices. By joining the corresponding points we obtain the group of straight lines that represent the Italic hour-lines.

To obtain the true time of sunrise or sunset on the days of the solstices (as well as for any other day), we can always use the formula, $\cos \mathrm{HA} = \tan\phi \tan \mathrm{D}$, where HA is the hour-angle of the sun at the instant of its rise or setting, ϕ the latitude, and D the declination.

Tables supplying this quantity for given values of ϕ and D are available, in some almanacs, especially nautical ones. Table 5 gives HA for the days of the solstices.

We can also use the information contained in ordinary almanacs

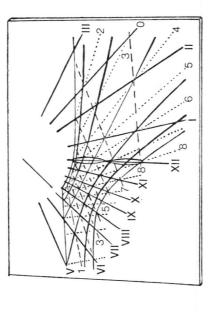

—— Standard Hours

– – – Babylonic Hours

········· Italic Hours

FIGURE 87   Perspective view of a dial carrying the Babylonic and Italic hour-lines. The numbering of the Italic lines is inverted. In this dial, the presence of these groups of lines is used to give immediately the times of sunrise and sunset as well as the length of the day for each day of the year

or calendars by taking the difference between the legal time for sunrise and sunset on a given day and dividing it by 2 to get the hour-angle sought. Yet, some care must be taken as to the latitude for which these special almanacs have been issued.

Remarks

We notice that on declining dials (like the one in figure 87 [see also Plate 10] which declines towards the SE), at sunrise the shadow of the tip of the style falls all the year over on the Babylonic hour-line zero. This line is horizontal and situated exactly at the height of the tip of the style. On a given day, the daily arc cuts this line at a point which defines exactly the hour-line of the sunrise on that day. In plotting the hour that marks this intersection from hour to hour along the daily arc, we obtain the points of successive Babylonic hour-lines. Proceeding in this way for the points of the daily arc of the solstice and then for the equatorial, we find a means of tracing Babylonic lines immediately and without calculations. This procedure cannot be used on this dial for the Italic lines since it is not possible to include the line of sunset. If we recall that the Babylonic lines and the Italic lines always intersect at the astronomical hour-lines of the dial, we may use the intersection of the latter with the Babylonic lines to trace the Italic hours. We can also use the fact (and this will be simpler) that the hours of sunrise and sunset are symmetrical with respect to the true noon.

Let us note too that, in general, on a vertical declining dial, there is no reason to extend the daily arcs to the left beyond the first Babylonic hour-line or to the right beyond the first Italic line, since the former is the line of sunrise while the latter is the line of sunset. If one of these lines is on the dial, it allows us, as we have just seen, to determine the time of the sunrise or the sunset on the date of the daily arcs that cut it; these data, moreover, afford a means of deducing the hours of the corresponding sunset or sun-

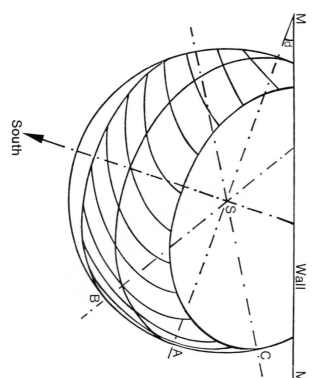

FIGURE 88

rise and thereby the length of the day. Usually, indeed, sunrise and sunset hours are not both simultaneously marked on a sun-dial in the way just told.

Finally, the fact that the intersections of the Babylonic lines with the Italic lines always lie on the astronomical hour-lines constitutes an excellent means of controlling the accuracy of the procedure used for their tracing.

B. *Tracing the Babylonic and Italic Lines with the Help of an Épure*

This drawing, as we have already noted, is a painstaking and time-consuming task. A reduced model of the celestial sphere which has the tip s of the style for its centre is placed in front of the dial (figure 88).

Since the same reasoning is followed for both Babylonic and Italic lines, we consider only the first.

In the figure, let us suppose that the observer is placed at the zenith of the tip of the style. He sees the plane of the horizon as the great circle in the plane of the figure, in which the dial is declining towards the SE. Between the summer and winter solstices, the sun rises in the sector of the horizon located inside the angle BSC. The portion BC of the horizon determines a plane with centre s which cuts the dial along a Babylonic line of order O. During the course of the day, the points B and C describe two circles on the celestial sphere – the tropics – which indicate the zone of the zodiac. Each point of the arc of a great circle BC describes along this zone an angle of 15°, 30°, and 45° around the axis of the poles, at intervals of 1, 2, 3 hours, etc. The respective planes determined by these arcs of circle and the centre s meet the plane of the dial along the Babylonic lines of order 1, 2, 3, etc. To determine one of these lines, it suffices to join two points of one of the arcs of a circle to

the tip s and then extend the straight lines so obtained as far as the dial, where their intersections determine the corresponding Babylonic lines. It is evident that we prefer to choose the two points either on the tropics or on one tropic and the equator.

In the following explanation, we have thus projected the segment of a great circle corresponding to the third Babylonic hour, i.e., the one that corresponds to a rotation of 45° of the segment BC around the line of the poles. We have done this same thing for the point A located on the equator, although it is not really necessary. The *épure* shows that the final projection creates three points on the dial which must lie on a straight line since it results from the intersection of two planes.

In projection I of the *épure* (figure 89), the observer is at infinity, on the line of the poles; in projection 2, he is on the diameter passing through the vernal point. The respective position of the two projections is determined by the angle ϕ of the latitude in 2.

The horizon is represented by the straight line HH′ in the latter projection. The straight lines EE′ and KK′ are the projections of the summer and winter solstices. In I we trace a circle with centre s′ and diameter EE′ = KK′, which represents the projection of the two tropics. The intersection of the horizon with the tropics in 2 determines the segment of horizon BC whose projection on the wall through the point s is being sought. The projection lines (parallel to s′s″) give in I the position of the segment B_2C_2 corresponding to B_1C_1 in 2.

At the instant of the third Babylonic hour, the segment has turned by $3 \times 15° = 45°$, around the axis of the poles (s′s″, in our case). Therefore in projection I we rotate the segment B_2C_2 (obtained from the lines originating at B_1C_1) by 45°, in the proper direction around s′. We draw its new projection in B′A′C′, an arc identical in shape with B_2C_2 and we look for its position in 2 with

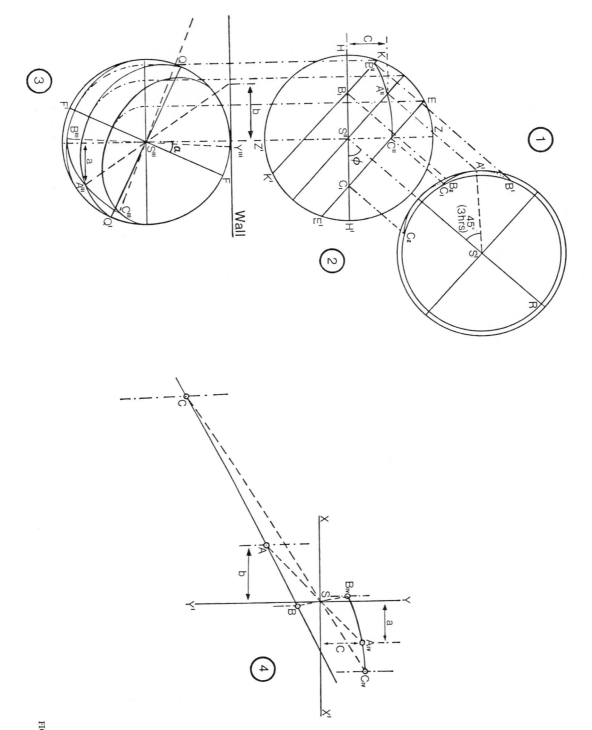

FIGURE 89

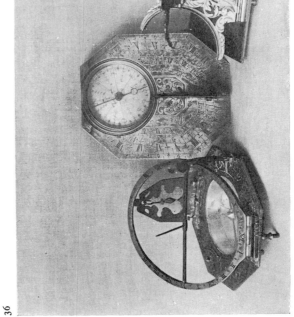

PLATE 35 In the centre, a horizontal dial with a pendulum for its setting and the daily arcs. On the left, a pocket equatorial dial. To the right, a pocket horizontal dial with its compass; the latter is a small navy compass, used almost exclusively on ships. (*Photograph: Musée de la Vie wallone, Liège*)

PLATE 36 Two equatorial dials of the eighteenth century. Note the pendulum for the setting of the one on the left. At the centre, the style of the equatorial dial is adjustable for the latitudes 43, 46, 49, and 52° N. The table carries the four corresponding dials. For the intermediate latitudes, one proceeds by interpolation. (*Photograph: Strasbourg Museum*)

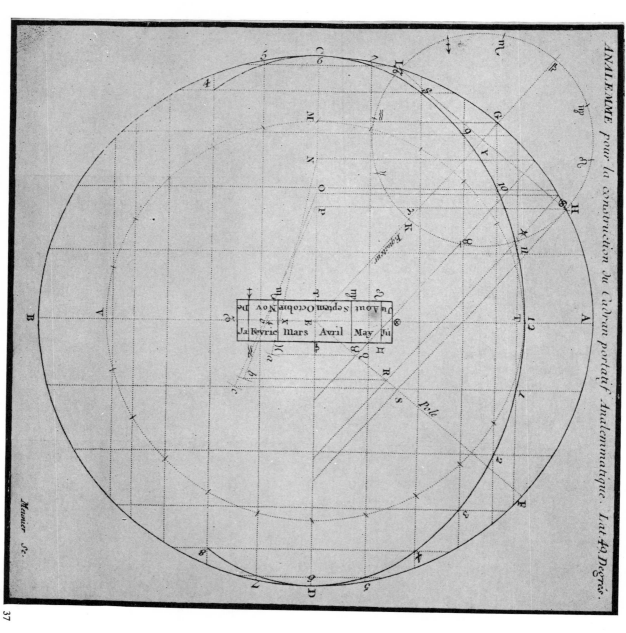

ANALEMME pour la construction du Cadran portatif Analemmatique. Lat 49 Degrés.

PLATE 39 Unfolded astronomical ring, according to an engraving found in the book of Bedos de Celle

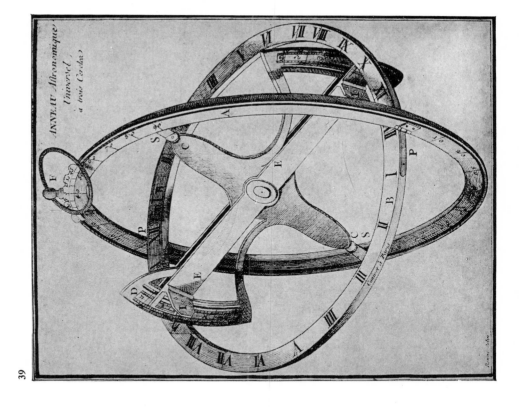

39

PLATE 38 Armillary sphere; it is very complete and it is meant for practical use, eighteenth century. (Engraving taken from the book of Bedos de Celle)

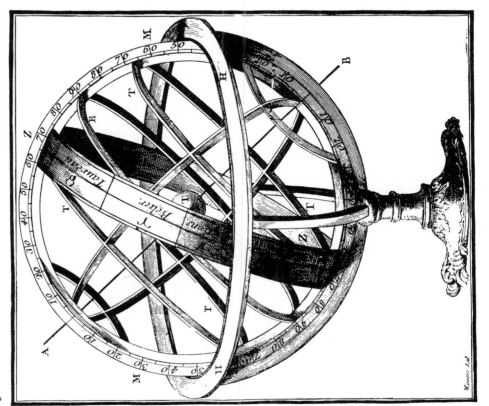

38

PLATE 40 The multi-faced dial of Mont Sainte-Odile in Alsace, first half of the eighteenth century. It is made up of twenty-four dials. On top we can distinguish the dial for local time, the only one which carries daily arcs. Above, the hours for Constantinople and Greece are indicated in two scripts. To the right, in the triangle, is the time for the "Congo-Africanum". The foot carries the arms of the prior of the Abbey of Neubourg (Alsace), where the dial was made. (*Photograph: the author*)

PLATE 41 Unfolding of the twenty-four dials of Mont Sainte-Odile. Some dials of the upper and septentrional faces are covered with moss; their deciphering was, therefore, not easy.

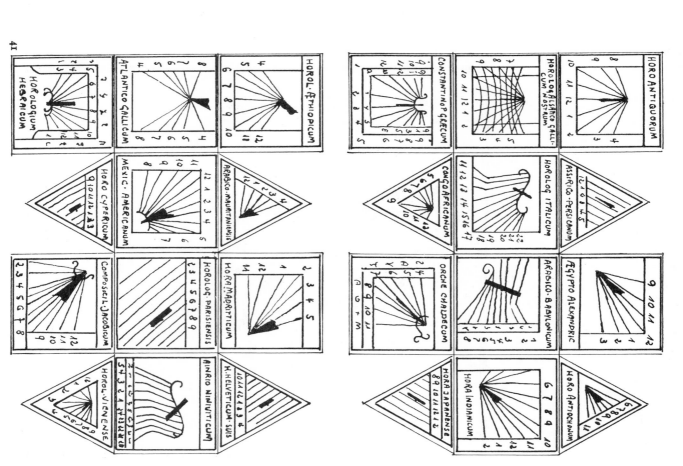

the help of the new projection lines. In this way B″A″C″ is obtained. Now consider a new projection in which the observer is at the zenith, as in figure 88 (which would result from projection 3 if the latter were performed for each hour of the day). The wall is drawn and then the circle of the horizon which is tangent to it. The intersection QQ′ of the equator with the horizon makes with this wall an angle d equal to the declination of the dial.

In order to find an arbitrary point located on a tropic or on the equator, we use projections 1 and 2 which supply all the necessary elements. For instance, the distance from A″ to ZZ′ in 2 corresponds to the distance from A″ to QQ′ in 3 and the distance from A′ to RR′ in 1 corresponds to that of A″ to FF′ in 3. Naturally we need some practice to visualize all these projections in three dimensions and to avoid mistakes. Similarly, we place the points B″ and C″ in 3 and we trace the curve B″A″C″, which is an elliptical segment, with the help of a suitable tool for drawing curves.

Now the intersection with the wall of a cone which has s″ for its tip and whose generating lines slide along the segment B″A″C″ must be found. This cone, as stated previously, becomes in fact a plane. This leads to projection 4, the image of the dial. s is chosen as the origin, the position of the orthogonal projection on the wall of the tip of the style, and we construct on it a co-ordinate system consisting of a vertical yy′ and a horizontal xx′.

In order to obtain 4, the observer must move to a location on the wall going through s. It is easy to find the perpendicular to the wall: we have its orthogonal projection of the point A‴ on the wall: we have its abscissa a in projection 2 (see figure 89) and its ordinate c in projection 2. This will be the point A$_{iv}$. We know too that in extending the straight line A‴s‴ to the wall in 3 we obtain the abscissa of the point A for which we are searching in 4. Therefore, we take an abscissa b in 4 and trace the straight line A$_{iv}$s which cuts the abscissa in A. Repeating the same operation for B and C, we note

(and this is a check) that the points A, B, and C lie on a straight line and the points B and C are respectively points on the daily arcs of the summer and winter solstices.

The preceding paragraphs have been concerned with the Babylonic day, starting at sunrise, and the Italic day, starting at sunset. The Italic system is still in use in the Jewish religion where the sabbath starts at sunset on Friday and ends with the sunset of the following day, following the old way of measuring time.

Only the daylight hours, as we know, were of use to the ancient peoples of the Mediterranean basin or to anybody else in those days. The time interval between sunrise and sunset was divided into 12 hours; if necessary, the night was also divided into 12 hours by a clepsydra. The hours of the day and the night could not have the same length, except during the equinox; neither could there be, strictly speaking, any equality between the hours of one day and those of the following day. These hours were therefore called temporary hours. They are also called biblical hours since it is assumed that the hours mentioned in the Bible were temporary hours.

Some mediaeval dials such as that shown in Plate 4 were designed to measure such hours. These dials, of which only a very few are left on some cathedrals, were installed facing the true south and consisted of a straight style and a table on which straight hour-lines divided a half-circle with a horizontal dial into 6 or 12 equal parts indicating intervals of time.

Evidently these hours were indicated by the shadow of the style, but they could not be equal to each other during the course of one day, a phenomenon that undoubtedly passed unnoticed by the majority of people and which could be checked only with an hour-glass. For precision, the lines should have been curves, but probably for centuries this error was of no importance. As we mentioned in chapter one in discussing the history of the sundial, the astronomers used equinoctial hours for their calculations, i.e., a constant unit equal to the twenty-fourth part of an equinoctial day. The only other persons interested in improved time measurement were the astrologers who assigned to each hour of the day the influence of a planet. Indeed, their influence on people since very earliest times should not be underestimated. We should not forget that an astrologer was present at the birth of Louis XIV to check on the precise moment of his entry into the world.

And so there still exist today in Europe some old sundials that carry on their face an extra division of the day into temporary hours; their hour-lines constitute, along with the seven classical daily arcs, the table of the 'masters of the day.' The table shown in figure 91 indicates in rectangular array the names of these 'masters' and of the planets which dominate the various hours of the days of the week. Such a dial, called a planetary dial, can still be seen on an inn in the village of Prutz in the Tyrol (Plate 14 and figure 90), the point of the shadow indicating the hour being still the tip of the style. We note here too that the draughtsman, imitating his mediaeval predecessors, contented himself with replacing the theoretical curves of the hour-lines by straight lines, committing, perhaps without his knowledge, a slight error. Note that the temporary line of 6 o'clock coincides with the 12-hour line of the classical dial and that the lines intersect in pairs on the line of the equinox, as they should do. Another most noteworthy dial of this type is chiselled on the walls of St Catherine Church at Oppenheim on the Rhine (Germany).

To construct a dial with temporary hours, one calculates the

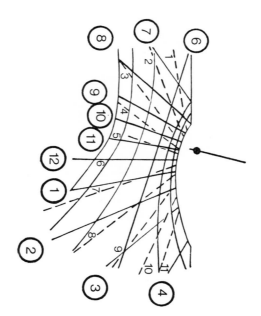

FIGURE 90 Dial bearing the temporary hour-lines indicated by dashed lines (Prutz, Tyrol). The lines of noon of the normal dial and of 6 o'clock (temporary) coincide

	Masters of the Day												Masters of the Night	
	1	2	3	4	5	6	7	8	9	10	11	12		
Sunday	☉	♀	☿	☾	♄	♃	♂	☉	♀	☿	☾	♄	♃	☿
Monday	☾	♄	♃	♂	☉	♀	☿	☾	♄	♃	♂	☉	♀	♃
Tuesday	♂	☉	♀	☿	☾	♄	♃	♂	☉	♀	☿	☾	♄	♀
Wednesday	☿	☾	♄	♃	♂	☉	♀	☿	☾	♄	♃	♂	☉	♄
Thursday	♃	♂	☉	♀	☿	☾	♄	♃	♂	☉	♀	☿	☾	☉
Friday	♀	☿	☾	♄	♃	♂	☉	♀	☿	☾	♄	♃	♂	☾
Saturday	♄	♃	♂	☉	♀	☿	☾	♄	♃	♂	☉	♀	☿	♂

Monthly Arcs	Zodiac	
Dec. 22	♑	
Jan. 20 - Nov. 22	♒	♐
Feb. 19 - Oct. 23	♓	♏
Mar. 21 - Sept. 23	♈	♎
Apr. 20 - Aug. 23	♉	♍
May 21 - July 23	♊	♌
June 21	♋	

☉ Sun ♀ Venus ☿ Mercury ☾ Moon ♄ Saturn ♃ Jupiter ♂ Mars

FIGURE 91 The astrologers assumed that each temporary hour was dominated by a planet which was its "master," the sun and the moon also being considered as planets. The days of the week still preserve in their names the memory of the masters of the planetary hours. On the planetary dial of Prutz (Plate 14) the hour-lines correspond to the vertical lines, the zodiac hyperbolas to the horizontal lines of the table, and the signs of the masters of the hour are inserted where space permits. (According to H. H. Kühnelt, *Die Sonnenuhren in Nordtirol*)

intervals of time between sunrise and sunset for the longest and shortest days of the year for the latitude chosen. These intervals are divided by 12, which gives the length of the temporary hours in standard hours for these two days. The addition, repeated 12 times, of the duration of the respective temporary hours to the times of sunrise on the two days under consideration gives two series of hours dividing the longest and the shortest days of the year into temporary hours. For example, let us assume that the sun rises at 3h 52m on June 21 at the place in question and that it sets at 8h 08m, the corresponding figures for the winter solstice being 7h 54m and 4h 06m true time. The length of the day is then 16h 16m on June 21 and 8h 12m on December 23. On June 21, the length of the temporary hour is (16h 16m)/12 = 1h 21m 20s; i.e., the temporary hours on that day start and end at

3h 52m, 5h 13m 20s, 6h 34m 40s, 7h 56 m, etc. (1)

On December 23, the situation is as follows:

Duration of the day: 8h 12m,
Duration of the temporary hour: (8h 12m)/12 = 41 minutes.

The temporary hours will start and end at

7h 54m, 8h 35m, 9h 16m, 9h 57m, etc. (2)

If we draw the hour-lines of series (1) on the daily arc of the summer solstice and those of series (2) on that of the winter solstice, using an auxiliary equatorial dial, we obtain two points on the solstices for the hour-line of each temporary hour. These points are joined by the intersection of the standard hour-lines with the line of the equinox. The dial of Plate 14 appears to have been drawn by following this procedure.

7 ELEVATION AND AZIMUTH OF THE SUN

There are in existence some rare dials that indicate the elevation and azimuth of the sun for each instant of the day. A sample of

this type of dial can be seen on the left in the group of three dials in Plate 18 (Cathedral of Strasbourg). In spite of the practical evidence supplied by these dials, we would rather believe that the prime reason for their installation was astrological – the dial of Strasbourg dates from 1572 – and that they were used to supply the basic information for the casting of horoscopes. We must note, however, that those three dials had been installed for calibrating the astronomical clock to be found in the cathedral at that time (and since replaced by another one).

The construction of these dials is simple since their tracing requires no astronomical information. The vertical plane of the table in figure 92, represented by its edge T, is supplied with a straight style of tip s and direction south, indicated by an arrow. On either side of this arrow angles of $10°$, $20°$, $30°$, etc. are drawn to indicate on the table the positions of the vertical lines corresponding to the azimuth of the sun.

The curves indicating the elevation of the sun are hyperbolas determined by the intersection of the table with cones of revolutions having s for their tip and a vertical axis. As usual, we rely on construction by points and projections. Let s be the projection on the dial of the tip s of the straight style (figure 93) and s its projection on the plane of the table. The length ss is the distance between the point s and the table. Let D be one of those vertical azimuth lines marked on the dial earlier. We look for the intersection of this line with the branch of the hyperbola indicating the height h of the sun above the horizon. The hypotenuse of the triangle $s'ss$ is in the horizontal plane going through s the distance from s to the straight line D, if we call s' the intersection of D with the plane. We project s and the plane defined by D and s around D by tracing an arc of circle of radius ss' around s', which gives on $s's$ the point s''. In s'' we trace, this time on the rotated plane, an angle $ss''p$ equal to h, one side of which gives the desired point P on D. By a single projection we thus obtain on P the points of intersection of all the curves which we wish to draw in making successively in *s'* angles of 10°, 20°, 30°, etc., as well as any desired intervening angles.

Proceeding with the other lines of azimuth as for D, we obtain other sets of points. Then the corresponding points need only be joined by hyperbolas to obtain the complete tracing of the trial. We shall study in chapter eight a special category of portable

dials, the so-called elevation dials, which are very convenient and were commonly used formerly; their construction implies knowledge of the elevation, the declination, or the hour-angle of the sun when two of these quantities are given.

The simultaneous presence on a dial of hour-lines, of daily arcs (lines of declination), and of the elevation lines just mentioned transforms it into an abacus which provides a sufficiently accurate solution for elevation dials (for further discussion, see p. 109).

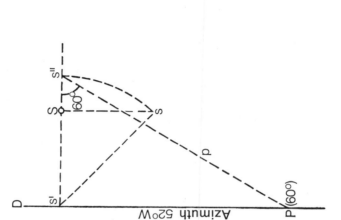

FIGURE 93 Construction of the intersection with D for the hyperbola for *h* = 60°

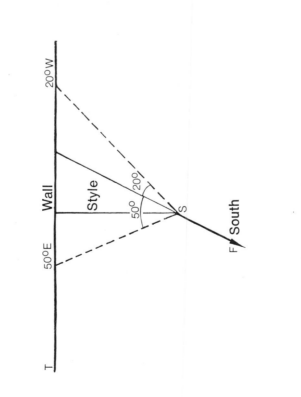

FIGURE 92 The azimuth of the sun on the dial. In order not to complicate the drawing, only the lines for the azimuths s 50 E and s 20 w of the sun have been drawn

CHAPTER SIX

Analemmatic Dials

I GENERAL REMARKS

In contrast to the classical dials with a fixed style parallel to the direction of the axis of the world, the analemmatic dials have a mobile style. This type of dial is not common except among the solar watches which we will study later. The analemmatic dial could not become popular because its style requires daily setting. This does not, however, interfere with its accuracy and in this respect it is the equal of classical dials. We do not know the name of the inventor of either of these dials. In the case of the classical dial, the originator must be commended for having created an instrument which gives pleasure to all sundial admirers when they come upon one on a wall or in a garden. The unknown inventor of the analemmatic dial, on the other hand, must have been a mathematician of some stature. The theory providing for such a dial was lost almost as soon as discovered but was revived by the astronomer Lalande in 1757; it must have been the work of a rather unusual creative genius. But this is only supposition because we have no real knowledge about the origin of the analemmatic dial. It is not impossible that this dial could have been discovered by chance or by trial and error during the Renaissance when a large number of people studied gnomonics and spent

much time on it. It is a fact that the craftsmen worked in a mechanical way from an *épure* in constructing it; this *épure* is found in most of the great treatises of the eighteenth century and Plate 37 shows a reproduction of it, taken from the famous book of Bedos de Celle. It is possible that some obscure mathematician, realizing how it performed, looked for and discovered the theoretical foundation; later, Lalande gave the definitive formulation of the theory which he published in the Memoirs of the Academy of Sciences in 1757. We must mention that Lalande's paper is very hard to find, except in some rare libraries. The author of this book has never seen it.

When we were surveying the history of the sundial, we touched briefly on the possible meaning of the Greek word which designates these dials. Vitruvius speaks in his book about 'people who are familiar with the analemma.' The theoretical conception of the analemmatic dial has many things in common with certain methods used by the Ancients to solve graphically some problems of astronomy, methods which their mediaeval successors replaced by the astrolabe. These graphical solutions are explained for instance in Ptolemy's 'Treatise of the Analemma' which dates from the second century AD. It is even conceivable that it was already familiar to astronomers in Hipparchus' time four centuries earlier.

The most celebrated analemmatic dial in France is the one found in front of the famous church of Brou in Bourg–en-Bresse. On the square in front of the church, there is a large ellipse formed by marble blocks on which the hours are carved at the appropriate positions in larger stones (figure 94). At the centre of the ellipse, a sculptured scale carries the names of the months located on the minor axis on which a vertical gnomon made of iron may be positioned, its shadow indicating the hours on the ellipse. The presence of a meridian in the shape of an 8 on the scale adds to the interest of the dial and supplies the equation of time without giving the correct time. We saw how this could be done on the classical dials. In fact, this meridian is purely documentary because its shape has meaning only between the hour–lines and the daily arcs of a classical dial. It cannot have been Lalande who put it there.

The ellipse is complete, probably for aesthetic reasons. It is evident that the shadow of the gnomon, or its extension, is displaced and supplies information only over a portion of the ellipse, never falling on that part where the figures indicate the hours of the night during the summer.

This dial dates supposedly from the time when the church of Brou was built (1506) and Lalande would have redrawn it in 1756. The dimensions of the ellipse are 11.18 metres by 9.09 metres.

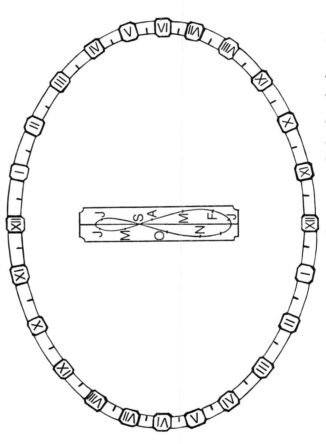

FIGURE 94   The analemmatic dial of Brou. The gnomon is displaced on the short axis of the ellipse and not on the meridian, whose presence here in the shape of an 8 is a mistake

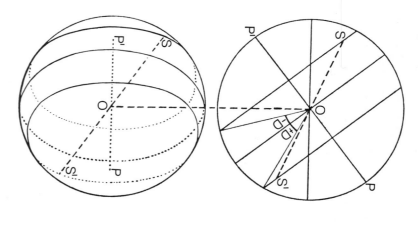

FIGURE 95

Incidentally, there is an identical dial at the back of a park in Dijon.

The analemmatic dial aroused interest because, in the eighteenth century, the high time of portable solar watches, craftsmen preferred to place a classical horizontal dial and an analemmatic dial on the same metallic plate (see p. 111). This double instrument could be accurately oriented in such a way that both dials indicated the same time, thus automatically giving the true time for the locality.

2 PROBLEM OF THE ANALEMMATIC DIAL

Let us consider the celestial sphere projected horizontally and vertically (figure 95) in conjunction with the daily circle followed by the sun while its declination is D and with a circle of declination of − D, and let us suppose that the centre of the sphere O is a shadow-casting object.

When the sun is in S, the shadow falls in S′, symmetric to S with respect to O. In the vertical projection, these circles are represented by the same number of ellipses and since the declination of the true sun varies continuously, there will be one for each day of the year. The envelope of all the ellipses is a great circle of the celestial sphere and, since the major axes all have the same direction, they will have various sizes while remaining similar to one another.

Before this idea is used for the construction of a sundial, some simplifications are considered. First, all the ellipses are assumed to have the same dimensions, and then the shadow-casting object is displaced in such a way that they coincide. It is logical to use the central ellipse, i.e., the projection of the equator and of the sun's trajectory at the equinox, as the standard ellipse with regard to size and to transfer all the other ellipses over it.

The distance by which the centre of one of the ellipses has to be displaced in

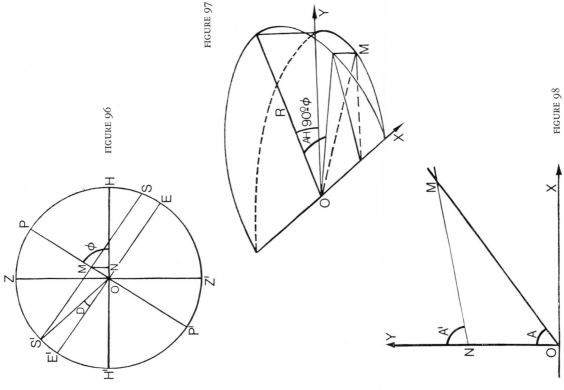

FIGURE 96

FIGURE 97

FIGURE 98

the vertical projection from that of the central ellipse is represented in figure 96 by the length ON, where we have:

OM = R sin D,

ON = OM cos φ.

Therefore

ON = R sin D cos φ,

if φ is the latitude of the locality.

This ellipse must now be stretched to the size of that of the equator. Since they are similar, the ratio of their dimensions is 1/cos D and we must take this into account in the displacement. Finally, the latter is given by multiplying ON by this factor, which gives:

Displacement = R tan D cos φ.   (1)

We see that it is positive in summer and negative in winter.

Figure 97 allows us to calculate the *x* and *y* co-ordinates of the points of the ellipse in a co-ordinate system in terms of R, φ, and HA; in the co-ordinate system, the *y* axis is the meridian, the *x* axis is the first vertical, the centre, O, the origin, and HA as usual designates the hour-angle of the sun.

We note that

x = R sin HA,

−*y* = R sin φ cos HA,   (2)

y being a negative quantity by reason of the contrary signs of latitude and considered ordinates in our figure.

If we displace the shadow-casting object along *y* as the declination D varies and if we use a gnomon, we shall see that at the equinox (D = 0), the shadow of the gnomon is the straight line OM, while for a declination D, it casts a line of shadow NM at the same true hour of the day (figure 98).

This brings us to an investigation of the nature of the geometrical locus of M, the intersection of the various lines of shadow, as shown in figure 98, and the dependence of this locus on the variables HA and D and on the fixed quantity φ.

For this purpose we use figure 28 which represents a celestial triangle and where A stands for the sun, at least for our purpose. This triangle gives

cos PZ cos ∠ ZPA = cot ∠ PA sin PZ − cot ∠ PZA sin ∠ ZPA,

i.e., if we call z the azimuth of the sun,

$$\cos(90° - \phi)\cos HA = \cot(90° - D)\sin(90° - \phi) - \cot z \sin HA$$

or

$$\cot z = \frac{\tan D \cos \phi}{\sin HA} - \frac{\sin \phi}{\tan HA}.$$

The equation of OM, the straight line which goes through the origin is

$$\cot z = -\sin \phi / \tan HA.$$

In figure 98 $\cot z$ measures the slope of NM. We obtain the slope OM by setting $D = 0$:

We also obtain the equation of NM by expressing its slope in two different ways. It is first given by (3); from analytic geometry it is also given by

$$\frac{x}{y} = -\frac{\sin \phi}{\tan HA}. \tag{4}$$

We obtain the slope OM by setting $D = 0$:

$$\frac{y - R \tan D \cos \phi}{x} \tag{5}$$

R $\tan D \cos \phi$ represents the displacement of the gnomon found in (1). The equation of MN is finally obtained by substituting (5) on the left-hand side of (3):

$$\frac{y - R \tan D \cos \phi}{x} = \frac{\tan D \cos \phi}{\sin HA} - \frac{\sin \phi}{\tan HA}. \tag{6}$$

Let us now look for the co-ordinates of the point M by using (4) and (6). From (4), we obtain

$$y = -\frac{x \sin \phi}{\tan HA}.$$

Replacing y by this value in (6), we get

---

or

$$\frac{x \sin \phi}{\tan HA} + R \tan D \cos \phi = \frac{x \sin \phi}{\tan HA} - \frac{x \tan D \cos \phi}{\sin HA} \tag{7}$$

in terms of the co-ordinates of the points M and N. The slope can then be written as

$$\frac{y_M - y_N}{x_M - x_N} \tag{3}$$

$$x = -R \sin HA.$$

In (4) this yields

$$y = R \sin \phi \cos HA.$$

We deduce

$$\sin HA = -x/R,$$
$$\cos HA = y/R \sin \phi.$$

Since the sum of the squares of the cosine and sine of the same angle equals 1, we get

$$\frac{x^2}{R^2} + \frac{y^2}{R^2 \sin^2 \phi} = 1, \tag{8}$$

i.e., the equation of an ellipse of major axis R and of minor axis R $\sin \phi$, which depends uniquely on ϕ and on the shape of which HA and D have no influence whatsoever as long as the displacement of the gnomon obeys rule (1). *Remark:* The equation of this ellipse, it will be noted, has already been shown in equation (2) obtained from the study of figure 97. Our calculation, which is more complicated since it starts from the celestial triangle on the table of the analemmatic dial, confirms the correctness of the result and underlines as well the fact that D and HA do not enter the equation.

3 CONSTRUCTION OF THE ANALEMMATIC DIAL

A. *Tracing of the Ellipse*

We have just seen that the ratio of the minor axis to the major axis equals $\sin \phi$. We may, therefore, either calculate the length of the minor axis from the major axis or construct it (figure 99)

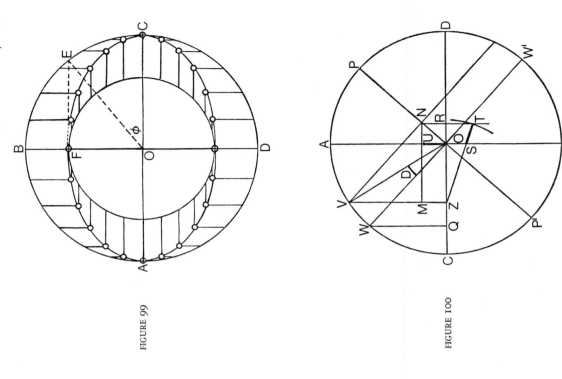

FIGURE 99

FIGURE 100

by dropping a perpendicular EF from E, OE being the upper side of an angle EOC = φ elevated on the radius OC. If *a* is the radius of the circle and *b* the length of the semi-minor axis, we could also trace the ellipse with the help of a thread once the distance 2*c* between the foci from the formula $c^2 = a^2 - b^2$ has been calculated. We recall that the length of the thread should be 2*a*, or, if it is a closed loop, 2(*a* + *c*).

A simple procedure consists in tracing two concentric circles of radii *a* and *b* and dividing each of them into 24 equal parts; the ellipse passes through the intersections of the horizontals from the inner circle and verticals from the outer circle, as shown in figure 99.

B. *Construction for the Location of the Gnomon*

This scale, as we have seen above, is related to the dimensions of the ellipse; it seems to have been mechanically constructed from the instructions of the books on gnomonics of the era. These instructions were always given without any theoretical explanations but an *épure* was always supplied similar to the one shown in Plate 37.

One can see from the scale where the gnomon was located on the central axis at the height of the month and, by interpolation, approximately on the date required.

It is probable that most, if not all, of the artisans never understood the basic reasons for the tracing expected of them. Perhaps they even knew by heart the successive steps to be followed. The fact remains that each line of the construction had to be there as we shall see in the proof to be given later. The refinement in the search for solutions inspires our respect for those obscure but certainly most able geometers who were the first to understand the true nature of the analemmatic dial, of which Lalande is supposed to have said that it posed the most subtle problems of gnomonics.

The construction started from the tracing of a circle of centre O whose diameter had the proposed length of the major axis of the ellipse. First the rectangular axis AB and CD were drawn, AB being the meridian, with origin in O (figure 100). Then the axis of the poles PP' were drawn, the angle POD being equal to ϕ, and finally the equator WW'.

Assuming that the search was for the point on the scale where the declination of the sun was D, an arc WV was drawn from W so that $\angle$ WOV = D. From V, the perpendiculars VZ and VN were dropped respectively to CD (the horizon) and on the axis of the poles; then from N, another perpendicular NR was dropped to CD. From W a perpendicular to CD gave the point Q.

An arc of a circle, having Z as its centre and QO as radius, intersected the extension of NR in T. The radius ZT of this arc of circle intersected AB (the line of the zenith) in S. ST was then the desired length which was measured from O along OA on the scale. The point U was obtained so that ST = OU.

Proof: The problem consists in displacing the gnomon to an extent that the two ellipses given by the projections of the circles having O and N as centres and OW and NV as radii coincide (see the vertical projection in figure 95). The two ellipses are similar but of different size.

The ratio of the sizes of the ellipses is VN/WO. Let us consider the two triangles VMN and WOQ. They are similar and in the ratio VN/WO of the circles and of the ellipses. If the ellipses were equal, one would need only to shift the smaller one by an amount OR. But to increase it in the ratio WO/VN (the inverse of VN/WO), a displacement is needed:

ST = OR. WO/VN.

But we have

$$\frac{ZT}{MN} = \frac{QO}{MN} = \frac{WO}{VN}$$

Therefore,

$$\frac{ZT}{MN} = \frac{ZT}{ZR} = \frac{ST}{OR} = \frac{WO}{VN}$$

We then have the relation (1) and the construction is correct.

4 THE ANALEMMATIC DIAL WITH THE HELP OF MATHEMATICS

Here the formula of page 103 will simplify everything.

If we choose 2R to be the length of the major axis of the ellipse, the minor axis has the length 2R sin ϕ. The scale of the displacement of the gnomon follows from (1):

Displacement = R tan D cos ϕ.

The points of the ellipse corresponding to the hours HA which we wish to indicate, are either obtained from the formulae (2)

$x = $ R sin HA,

$y = $ R sin ϕ cos HA,

or by drawing the ray which makes an angle α with the minor axis, so that

tan $\alpha = x/y = $ tan HA/sin ϕ.

This formula shows that the hour-marks indicated on the ellipse are symmetrical with respect to the axes. We need to compute them in one quadrant only.

(1)

CHAPTER SEVEN

Moon Dials

On moonlight nights, especially when the moon is full, it is natural to think of using the often very sharp shadow of the style on the dial for the reading of time. This idea is very old and completely justifiable.

It will be recalled that figure 23 showed phases of the moon in the exterior circle as observed from the earth for various stages of lunation, which, as we recall, extends over $29\frac{1}{2}$ days. The phases of the moon are marked by the value of the angle which the moon makes with the sun, i.e., the difference between the right ascensions of these two bodies. At the instants of new and full moon, the angle is o and 12 hours, respectively. If one watches the sundial at the precise instant of full moon, the time indicated is the true time. Similarly at the instant of new moon, the sun and the moon having the same hour-angle, their common shadow on the dial (if the moon could produce one), would indicate the same time, i.e., the true time.

When the moon has moved away from the sun by 15° of right ascension (one hour in units of time), the reading of the lunar shadow has to be corrected by one hour in order to obtain the true time. For 30°, the correction is 2 hours, and so on. Finally, for a difference of 180°, namely, when the moon has gone from new to full, the correction is 12 hours – which is, of course, unnecessary

since the reading on the dial gives the true time of the night without correction.

The difference between the right ascensions of the sun and of the moon has been called the *lunar angle* and one may write:

True time = time indicated by the lunar shadow on the dial + the lunar angle expressed in hours.

But how can we know the value of this angle without an almanac for the year?

The various phases of the moon, as stated above, characterize the epochs of its lunation, in other words, what we have called the age of the moon (p. 33). With practice it is possible to determine this age from the corresponding appearance of the moon shown in figure 24. This procedure is used since it is the simplest and requires no apparatus. It is easy to make a mistake unless one recalls the date of the last lunation, and this is not often the case. Compounding the possibility of error, the originators of some dials have simplified the problem by rounding off the duration of the lunation to 30 days, giving on their dials a table indicating for each day the value of the lunar angle to be added to the reading.

There is such a dial at Queens College in Cambridge (Plate 9) which owes part of its fame to this detail. For each interval of fifteen days between the new and the full moon, and between the full moon and the next new moon, its table shows the value of the lunar angle, i.e., the correction to be added to the hour indicated by the shadow, if it is sharp enough to be read. This condition must have induced the constructors of lunar dials to locate the table of corrections at the bottom of the dial, i.e., as near as possible to the observer, to use a white background, and large dimensions as compared to the surface of the dial itself.

There is another way, not quite so spectacular, of indicating the value of the lunar angle, i.e., by placing at the bottom of the dial a small graph similar to that shown in figure 101. But the main problem is not yet solved, since the age of the moon must still be deduced from its phase, unless this can be arrived at by other means. On the graph, the indications between the first and last quarters have been omitted since experience shows that even on a clear night the lunar light is still too weak to throw a sufficiently defined shadow.

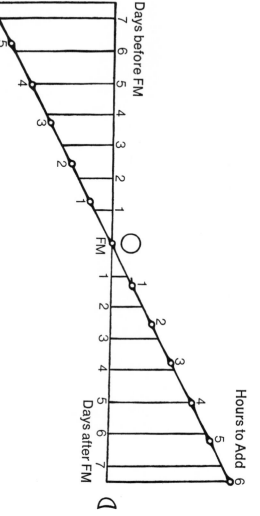

D
Days before FM

7 6 5 4 3 2 1 FM 1 2 3 4 5 6 7

4
5

Hours to Subtract

Hours to Add

6
5

Days after FM

C

FIGURE 101   According to whether it is before or after the full moon, the hours are subtracted from or added to the indication given by the dial. In the graph the lunation has been estimated to last 30 days. Using a graph for a lunation of 29½ days would not improve the result very much

Once an estimate of the age is obtained, the length of time to the full moon is deduced, a length that varies between $-7\frac{1}{2}$ and $+7\frac{1}{2}$ days. The length is read on the horizontal scale of the graph and the vertical raised at this point gives, by interpolation on the oblique scale, the value of the lunar angle falling between -6 and $+6$ hours. This correction will be 48 minutes for every day.

Of course, in order to obtain the mean time, one must add to the result the corrections for the longitude and the equation of time. Perhaps this is asking too much from people who walk by night!

Moon dials are scarce for the very good reason that they are not easy to use and that, basically, the result, always more or less compromised by the inaccuracy of the reading and the poor estimate of the age of the moon, may show a considerable discrepancy with real time. So one may assume that the constructors of lunar dials meant them mainly for those to whom time is of little importance. It is said that lovers fall in this category.

CHAPTER EIGHT

Elevation Dials, Azimuth Dials, Direction Dials, and Others

1 ELEVATION DIALS

The purpose of the dials to be studied in this paragraph is not to measure the hour-angle of the sun or its azimuth, but rather its elevation above the horizon. Since the sun goes through the same elevation twice in one day, the observer must be aware of the part of the day, morning or afternoon, during which he takes the reading.

In the celestial triangle, there is a relation between the hour HA, the latitude ϕ, the sun's declination D, and its elevation h, which is given by the well-known mariners' formula:

$$\cos(90° - h) = \cos(90° - \phi)\cos(90° - D) + \sin(90° - \phi)$$
$$\sin(90° - D)\cos HA$$

or

$$\cos h = \sin\phi \sin D + \cos\phi \cos D \cos HA.$$

Construction is based on this relation.

It is quite probable that the artisans of past days did not know this equation under its explicit mathematical form and preferred to use tables, some of which are found in old treatises. We must remark that now highly accurate tables of the same type are still

in use on ships and especially on aircraft where the time required for the calculation of a point must be minimal.

The reader should be aware of the fact that a dial carrying on its table the hour-lines of true time, the daily arcs for a certain number of days, and the line of elevation of the sun may be used as an abacus to calculate any of the three elements h, D, HA, when two of them are given (under the latitude for which the dial has been set). In the chapter on moon dials, we mentioned the dial of Queens' College in Cambridge which fulfils these conditions. It is very old – dating from 1733 – and others like it must have been in existence. The knowledge required for the construction of each of the family of lines carried was normally accessible to the makers of dials, so we may assume that this simple and practical abacus was used for the determination of the lines traced on elevation dials.

As in all other types of dials, the possibilities for the construction of elevation dials are innumerable. We shall, therefore, describe only some basic types, to which the great variety of dials to be found in museums and private collections may be reduced.

The first type consisted of a rectangular plate ABCD with its sides approximately in the ratio of 2 to 1. From the point C, the side BC carried over two-thirds of its length a graduation of 18 divisions,

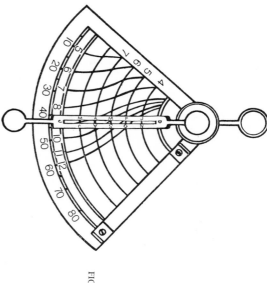

FIGURE 102

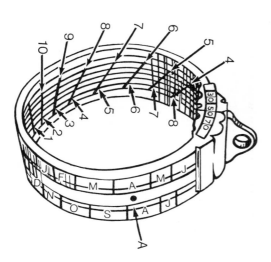

FIGURE 103 The point of suspension could be altered according to latitudes from 30° to 70°.

each corresponding to 10 days of the year, and extended over the overlapping intervals from December 21 to June 21 and from June 21 to December 21 (Plate 25). Circles were traced with B as their centre, one for each notch on the scale; in addition, a plumb-line was fixed in B, carrying a mobile weight which could be stopped at the level of the date indicated by any one of these circles.

At A and B two small bored stems, were fixed, which could be oriented in the direction of the sun so that its rays going through B would fall on A. The smaller sides of the rectangle then had an inclination equal to the direction of the sun, when the rectangle was held in a vertical plane.

For a given elevation of the sun, the time of the day varies depending on the date, i.e., according to the value of its declination; therefore, curves in the shape of an s and serving as an abacus, were drawn point by point on the dial. Only the curves for the whole hours were traced in this way. By interpolation, the dial could be used to determine the time throughout the year. It was sufficient to move the lead weight of the plumb-line to the date of observation, to orient the plane of the dial vertically, and to bring into A the rays going through B in order to read on the curves or by interpolation within the intervals the hour indicated by the lead weight.

To trace the curves with the help of tables or by any other means, a graduated arc of a circle of centre B (shown on Plate 25), which was removed once the curves were drawn, was used. Figure 102 shows a different but less awkward variant of the same dial.

Our second example is probably the commonest of all the elevation dials. Many so-called *shepherd's dials* can still be found in all museums and in almost all private collections.

The shepherd's dial consists of a cylinder capped by a movable top to which there is attached a style that could be folded along the cylinder while it was being carried about.

disc with all the ornaments required by the taste of the era or perhaps by the need to increase sales.

Examples of the application of the elevation of the sun for the determination of the time could be multiplied endlessly and their description would fill many chapters. Every museum offers surprises in this field and often the problems raised regarding their principle and their functioning cannot be easily resolved even by experts. When the sundial was at its peak, investigators and inventors were numerous and apparently had lots of time at their disposal.

2 AZIMUTH DIALS

The other important type of dial to be mentioned is the one in which time is determined from the variation of the azimuth of the sun, i.e., from the angular distance between the vertical which passes through the sun and its meridian.

We have already met the analemmatic dial which belongs to this category and was discussed in chapter six; it is, in fact, practically the only representative of azimuth dials. As we have seen, the construction of this dial was much more subtle than that of any of its relatives, but its use with a small horizontal dial on the same plate had the great advantage of not needing a compass for orientation. Plate 33 illustrates such a dial. In order to orient the instrument in the meridian, it was sufficient to rotate it till the shadows of the style and of the gnomon on each dial indicated the same time. This was an almost instantaneous process.

The wide dispersion of the analemmatic dial with portable dials is due essentially to this valuable function which resulted from the different principles governing the two dials. We must note, however, that it could be used only in the latitude for which it had

Since the instrument had to be held vertically when in use, a thread was fixed in the tip of the cap (Plate 24). In this position, the style was oriented in the direction of the sun. The hour should then be indicated by the shadow on a calibrated curve of the cylinder. In order to make it usable for various declinations of the sun, the dates of the year were traced around the cylinder at intervals of 10 days, taking into account the overlap at the times of the year when the declination of the sun was the same. The inscription of the elevations of the sun for the various hours gave rise to continuous curves analogous to those of the previous dial. Plate 27 shows the unfolding of the surface of the cylinder of such a dial with the lines carried. This plate as well as the previous one has been taken from the treatise of Dom Bedos de Delle (1760).

Understandably, the simplicity of the shepherd's dial has given rise to many variations, such as that shown in Plate 26 which has the appearance of a rigid flag, usually made of metal and installed on a movable support to ensure a vertical setting. Instead of turning around the axis of the cylinder, the style is displaced along the upper edge of the dial and when in use it is directed towards the sun; the plane of the dial stands vertically as was the case with the shepherd's dial and must be oriented in a direction perpendicular to that of the sun's rays, i.e., to the azimuth of the latter.

The last great variation of these dials, a ring, follows the same principle. Here the style is replaced by a small opening, A, in the surface of the ring which itself hangs freely from a small loop (figure 103). Once the ring is properly oriented, the light of the sun crossing the opening falls on the interior surface of the opposite side where a smaller and distorted drawing, substantially the same as the one shown in Plate 26, is reproduced. The smaller dimensions of the instrument made the time reading slightly less accurate.

The same dial was sometimes made in the shape of a metallic

been devised. But this is equally true for the horizontal dial of the instrument shown in our plate.

We know that the gnomon of an analemmatic dial had to be shifted according to the date. So it is not surprising that somebody thought of keeping the gnomon fixed and of making the ellipse movable.

Indeed Plate 33 also shows an instrument made of ivory in the shape of a diptych which, when opened, disclosed a circular horizontal dial arranged around a hemispherical cavity; the style was represented by a thread whose inclination evidently had to be calculated in terms of latitude. As we have just seen, the instrument had to carry an analemmatic dial also in the style of the era. We can see its mobile ellipse at the bottom of the cavity where it could be shifted in a direction perpendicular to the axis of the hinges. In an excess of luxury, the gnomon was a sort of cap in the shape of a paraboloid topping a magnetic needle, which, as we now know, was superfluous. Indeed, the orientation of the instrument based on the indication of the magnetized needle could only falsify the reading because of the magnetic declination. But this declination was just beginning to attract attention then.

Similar instruments were worked out carrying an automatic movement for the displacement of the ellipse with the help of a cammed wheel on which the dates were engraved. The wheel jutted slightly beyond the edge of the box and it was turned until the date of the observation could be seen in a small window. The ellipse, pushed by the cam, then lay in the proper position with respect to the gnomon.

Other azimuth dials have been devised with a fixed gnomon. Strictly speaking, the obelisks of antiquity could have been such dials. Indeed, it served to mark on the ground throughout the year the tip of the shadow on the various whole hours of the day. The loci of the points so obtained would give hour-lines in the form of curves. Such dials were, in fact, constructed.

3 DIRECTION DIALS AND COMPOSITE DIALS

In this paragraph, we return to the classical dial with its style fixed on the table and oriented parallel to the axis of the world. There is such a vast range of different varieties of these dials, however, that we can say something only about the main types, the variants of which appear in the most unexpected shapes.

For all these instruments, also called solar watches, the masters of the goldsmith's craft and of engraving, of precision mechanics, and of compass-making pitted all their art and ingenuity against one another to attract and satisfy a large clientèle whose tastes varied with the times. There is no doubt that such dials were found in every house at one time or another. The wealth of the new ideas is reflected in the diversity of the creations which took, among other forms, the shape of boxes, cubes, large and small rings, canes, books, armillary spheres, and even boats.

This book could not give an even approximate description of their various forms. The interested reader will find many examples of them in certain museums, especially in the Musée de la Vie wallonne in Liège, from which many of our plates were obtained and whose exceptionally rich collection was bequeathed by a collector of Nordic origin, Max Elskamp, a poet and lover of the folklore of his country. Thanks to his legacy, the collection of dials in the Walloon museum is probably one of the most important in the world. In the United States a fine collection is shown at Harvard University Observatory.

The type of dial found most frequently in those collections is the horizontal dial. As an example, we have chosen an item from the museum of Liège which appears in the shape of a diptych and

PLATE 42   Four-faced dial set on top of a
column at the intersection of two roads in
Alsace, eighteenth century
(*Photograph: J. Baumann*)

PLATE 43   Small modern meridian on the
Church of Urmatt (Alsace). In Alsace,
many churches have similar meridians to
allow the setting of the church-clock,
nineteenth century. (*Photograph: the author*)

PLATE 44   Dial in the shape of a sphere
carrying the date 1732. It carries the
meridians for every 15°, therefore 24 in
all. The accuracy of its reading falls very
much behind that of the flat dials.
(*Photograph: the author*)

44

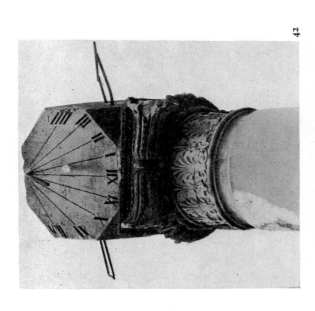

42

43

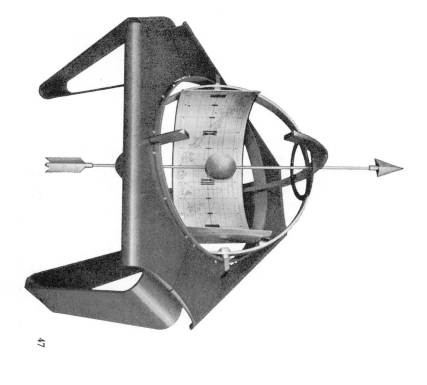

PLATE 45   The southern face of Brother Arsène's gnomon at Annecy. The stand of the star carries dials oriented according to the points of the compass; two equatorial dials for the winter cover the two uppers rays of the star. The southern face of the base carries a dial for a special type of information. (*Photograph: A. Gobeli, Annecy*)

PLATE 46   The northern face of the same dial. The summer equatorial may be seen on the upper face of the star, which carries polar dials between its rays. The stand of the star carries a small septentrional dial. (*Photograph: A. Gobeli, Annecy*)

PLATE 47   Modern equatorial dial made in Switzerland. It is the model adopted by a number of large cities for the ornamentation of the public promenades. The accuracy of the reading is of the order of one minute. (*Photograph: Gübelin, Lucerne*)

47

46

45

48

50

49

PLATE 48 Horizontal dial made of sandstone in the Botanical Garden at Bremen. The signs of the zodiac are purely decorative. (*Photograph: Gartenbauamt, Bremen*)

PLATE 49 Detail of the dial in the Botanical Garden at Bremen (*Photograph: Gartenbauamt, Bremen*)

PLATE 50 The original presentation of the correction of the equation of time of the dial in the Botanical Garden at Bremen. The upper and lower inscriptions mean respectively "the dial is slow" and "the dial is fast". (*Photograph: Gartenbauamt, Bremen*)

PLATE 51 The artistic and monumental armillary sphere of Phillips Academy, Andover, Massachusetts. The equatorial dial is buried by the large number and the size of the accessory symbols. (*Photograph kindly supplied by the Phillips Academy*)

is the work of a craftsman from Nuremberg (Plate 28). This seventeenth-century dial is made of ivory and is worth examining closely for the numerous markings carried on it. Indeed, the hour-lines of the dial calculated for the latitude of Nuremberg may be distinguished on the frame of the horizontal portion. Concentric circles around a hemispherical cavity contain other groups of hour-lines for latitudes varying between 39 and 51°. The thread which serves as a style is fixed to a vertical scale of latitude on the cover, which also carries a list of thirty-three European cities and the latitude of each to allow a more general use of the instrument.

To the left of the scale the shadow of a small gnomon draws daily arcs; only those for which the length of the day consists of the total number of hours have been traced on the dial. The figures go from 8 to 16. The text, in Old German, reads: "The tip indi-cates the twelve signs and the length of the day," On the right, a dial of temporary hour-lines is found under another gnomon be-tween the daily arcs of the solstices. It is accompanied by the words: "The tip indicates the planetary hours," here meaning temporary hours.

In the spherical cavity of the diptych, a vertical gnomon gives the Italic and Babylonic hours; the instrument also carries a com-pass.

Other horizontal dials are shown in Plates 29 (upper portion), 30 (top centre), 33, 34, 35, and 36 (centre). Very often the style may be adjusted for various latitudes. The modifications in the lay-out of the dial implied by these displacements are not always allowed for, as was the case for the previous dial. In order to limit the error, the variations in latitude were kept within a narrow range. In any event, as we have said before, errors in the movement of time were relatively unimportant in those days. Usually, the horizontal dials carried a compass, or an analemmatic dial, which must have been appreciated, especially after the discovery of

magnetic declination (Plate 33). It has been suggested, and perhaps with reason, that the gnomonists of that era were the first to notice a difference between the true north and the magnetic north.

Sometimes a horizontal dial and vertical dials which were meri-dional, septentrional, occidental, and oriental were put together on a wooden cube (Plate 29). Generally the tables of these dials were printed on paper and glued to the wood. These more spec-tacular but less accurate dials were often provided with a compass and a plumb-line for setting.

When discussing the equatorial dial, we saw that it was highly valued for the regularity of its divisions. These dials appeared in the past in many forms, nearly always in yellow metal, but some-times gilt. Their mechanical precision was equalled only by the fineness of their engraving. Plate 31 shows a small instrument of this type folded for carrying. On the left, the hinge of the equa-torial circle which carries the style folded in its plane can be identi-fied. On top, a dial for the latitude could be pulled out around a hinge which is partially visible. Such a dial was evidently uni-versal and could be used anywhere in the northern hemisphere. One needed only to reverse the order of the hours in order to make it usable all over the world.

Variants of this instrument can be seen in Plates 30 and 35, some of them carrying compasses. If the reader examines these instru-ments in detail, he will appreciate the precision work, the creative imagination, and the artistic sense which have been put into their conception and fabrication.

Similar instruments, devised to serve in any latitude, are still for sale in simplified form. There is no doubt that they were useful to explorers and pioneers. Plate 32 shows a small instrument, folded for carrying, which is known as an *astronomical ring*. It is a modi-fication of the equatorial dial and is found in a great variety of forms in museums. The dial shown in Plate 39, taken from the work of Bedos de Celle of which we have already spoken, is an example. Here the instrument is open and ready for use. The vertical great circle is suspended in F by a small ring. The instrument must be built in such a way that its centre of gravity coincides with the centre of the circle. By sliding it along the vertical great circle, the support of the suspension ring F may be placed facing the latitude indicated by a graduated scale on the circle. At the points of origin of this graduation, another great circle is attached, vertical to the first one. When in use, the vertical circle is placed in the plane of the meridian so that the second circle represents the equator.

A mobile pivot joins the points of the vertical circle of latitude 90° and represents the line of the poles. The pivot supports two wings which end in arcs of circles on which are engraved the graduations of the sun's declination and the signs of the zodiac. An axis crosses the centre of the pivot and supports a pinnule which is directed towards the sun.

When the vertical circle is oriented in the plane of the meridian, with the suspension ring set for the latitude of the locality and the viewer facing the sun, the wings of the central pivot indicate the declination of the sun and the sign of the zodiac in which the latter is located. The astronomical ring was an instrument with world-wide application – for use in any latitude. Unfortunately, the great mobility of its suspension must have made its operation difficult. Nevertheless, it has taken thousands of forms, so different that each piece exhibited in the museums appears to be unique in its category.

It was, perhaps, the instability of the suspended position of the astronomical ring which led to the concept of dials in the shape of a sphere. This rather uncommon dial looks like a school geographical globe. The inclination of the axis of the poles had to correspond to the latitude and it had to be parallel to the axis of the

world. The equator was divided into 24 equal parts. The sunlight separated the sphere into a clear and dark hemisphere. The line of demarcation indicated on the equator hours 12 hours apart whose half-sum gave the true time. To eliminate this calculation, the instrument was sometimes supplied with a semicircular knife edge made of wood which rotated around the axis of the poles. When it turned until its shadow on the sphere became as narrow as possible, it indicated on the equator the hour-angle of the sun, i.e., the time (figure 104).

Plate 44 shows a stone sphere, 48 centimetres in diameter and dating from 1732, on which the meridians are drawn at intervals of 15°. This very heavy dial was installed permanently in a garden.

There are various polyhedral dials, always interesting but of subtle construction, which are not equatorial dials. We shall mention an example in the next chapter.

Another simple dial consisted of providing a horizontal rod with marks corresponding to the hour-lines of a polar dial and then setting it perpendicular to the meridian; a style was then placed at the proper distance. Because of its clumsiness, this dial is not common.

These few examples can convey only a very remote idea of the richness of the gnomonic production of the seventeenth and eighteenth centuries, which is reflected by only a few displays in the showcases of museums.

One final, and curious, example is the dial shown in Plate 34 which was meant to be used at a distance and to indicate to a scattered population the exact instant of noon by a cannon shot. It recalls the procedure formerly followed in seaports to indicate the exact instant of mean Greenwich noon to the various ships in harbour: as an aid in checking the chronometers on board, a cannon was fired while a balloon fell from the mast of a semaphore. This noisy and extraordinary type of dial was far from rare.

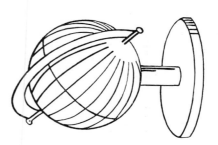

FIGURE 104   The spherical dial. The shadow of the movable semicircle should be as thin as possible

CHAPTER NINE

Remarkable Dials – Past and Present

In chapter one, we reviewed the history of the sundial. Now, to complement that chapter, we shall attempt to illustrate the history of the sundial with examples from the past and with others of more recent origin. Dials are no longer found in the workshop of the artisan but rather in public squares, on the façades of buildings, or even inside buildings which are open to the world. Our selection of examples, which is limited by necessity, extends over many countries, but the reader should not be surprised at the author's preference for dials located in his own area. If the city of Strasbourg, because of its cathedral, appears more often than seems just, the author offers the excuse that it is his place of birth and that its cathedral is gnomonically richer than the others which he could inspect in Europe.

At the beginning of this book we reviewed the rare famous dials of antiquity. Among these is the altar of King Ahaz, which is mentioned many times in the Bible but whose details are unknown. We have spoken also of the Tower of the Winds in Athens, reproduced in Plates 2 and 3, the old dials of which are still recognizable.

In books on gnomonics, numerous references are made to a dial (1389) on the old Paris city hall, with temporary hours and held by an angel. This type of dial must have been frequent on the

ancient cathedrals because we still possess a rather well-preserved example of it on the cathedral of Chartres (Plate 6).

The left support of the south transept of the cathedral of Strasbourg carries another dial, the Adolescent with the Sundial (Plate 4). Close to it, on the façade of the transept, a famous composition depicting a man with a turban under an ogive of carved grapes bending over a dial is called the *Man with the Sundial* or the *Astrologer with the Sundial* (Plate 7). This dial bears the date 1493. It declines by 29° 40′ to the southeast, and examination of its hour-lines shows that it is properly drawn for an inclined style making an angle of 41° 30′ with the vertical, the complement of the latitude of Strasbourg. It is, therefore, a modern dial with equal hours. Since there is no reason to doubt the authenticity of the date and since no older dial with an inclined style is known either in France or in Germany, it could very well be the first, or one of the first, sundials with equal hours. The transition from symmetrical dials with temporary hours to assymetrical dials seems to have embarrassed its constructors to the point that they decorated it with superfluous lines for the late afternoon hours that are never reached by the shadow of the style because of the declination of the dial. This was undoubtedly a concession to well-established habits. The year 1493 could then mark an advance in the history of gnomonics, the appearance on a public monument of the first known dial with equal hours.

Leaving Chartres and Strasbourg, let us look at the Ponte Vecchio in Florence where a curious and probably older dial (Plate 23) is found on a corner of the dilapidated roof of a building bearing an old coat of arms – rather mishandled by the capricious improvisations of the Tuscan masons. A glance suggests that it was set in its present place by a lover of curiosities. It is difficult to study it closely since it is almost inaccessible, but without doubt it is the most ancient of the known dials with temporary hours. Its style is directed towards the south and it lies approximately in the plane of the meridian. The half-circle of which it is made up is roughly divided into four equal parts between which traces of a subdivision can be distinguished. Its origin is unknown.

While we are in Florence, we recall that there is a little hole for letting through the rays of the sun at noon inside the nave of the celebrated cupola of Brunelleschi in the cathedral of that city; the hole is very close to the top and on the meridional side (Plate 17). On the ground, 106 metres below, the astronomer Toscanelli who devised this gnomonic installation, drew a meridian, i.e., a straight line located in the meridian of the luminous hole over which the passage of the rays indicates the exact true noon. Since the declination of the sun varies, diverse markings have been entered along this meridian, allowing recognition throughout the year of the passage of the sun through the signs of the zodiac. This is one of those monumental installations for the determination of the hour of noon about which we have already spoken and which obviously gain accuracy with the height of the gnomon (here the opening in the cupola).

Such installations were found at one time in all cities. The cathedral of Strasbourg is no exception. In the wooden door leading into the famous astronomical clock, a little circular opening has

been gouged out which corresponds to that in the cupola of Florence and the meridian lies inside at a distance of two metres on a vertical board duly marked with the limits of the seasons and the signs of the zodiac.

The meridian, large or small, must have played an important role in the determination of the exact time until quite recently. Indeed, we find them in use all over the French province of Alsace for the setting of church clocks similar in form to the one shown in Plate 43. Let us recall what we said at the end of chapter one about the regulation of the clocks in the railway stations of France with the help of precision heliochronometers which were used until the beginning of the present century (figure 12).

The group of three sundials shown in Plate 18, which decline by 29° 40' to the SE, is said to have been carved on the face of the southern cross window of the cathedral of Strasbourg around 1572. The upper dial indicates the hours in Roman numerals written in the Gothic characters then in use since the Middle Ages. To the left and below, we find one of those rare dials on which a straight style, invisible on our plate, throws its shadow on groups of vertical straight lines indicating the azimuth of the sun and on hyperbolas determining its elevation. This type of dial has been discussed in chapter five. Since the dial on the right is also equipped with a straight style it indicates the Babylonic and the Italic hours. Because of its height above the Place du Chateau, the group is difficult to photograph. Our plate gives the reproduction of a seventeenth century engraving in which, incidentally, some architectural details have been simplified.

To continue our review, a new idea was elaborated in Basel in 1731. It concerns a small octagonal stained window 14 cm across made from a small sketch attributed to the painter Jean-Rodolphe Huber. This window is now one of the gnomonical treasures of the historical museum of that city (Plate 8).

The inventors' notion was to place the window in the opening of a properly oriented wall with the style outside. The numbering was consequently inverted so that the dial, as it appears on the plate, seems to be meant for use in the southern hemisphere, where the numbers would effectively increase clockwise. It appears from the position of the hour-lines, assuming that the centre of the dial lies at the rather ill-defined intersection of the extensions of the short segments of hour-lines above the numbers (probably they are not properly drawn), that the dial was inclined in its original setting. The inscription (motto) is typical of those days. It is in old German and means *Time flies, death approaches. Man, think of that and fear God.*

There was a time when it was customary to install on the faces of a polyhedron, especially in convents, many dials which gave simultaneously the true time for many countries. The well-known convent of Mont Sainte-Odile in Alsace has such a multiple dial on one of its terraces (Plate 40 shows its southern side). The polyhedron of red sandstone on which it lies carries 24 dials in all, among which 8 are triangular in shape. We find a dial for the local time – HOROLOGIUM ALSATICO GALLICUM NOSTRUM – the only one carrying the usual seven daily arcs and also dials of temporary, Babylonic, and Italic hours. Others indicate the time in Paris, Vienna, and Madrid. Many refer to regions that are more or less vaguely delineated, Ethiopia (Africa?), Arabo-Mauritania, Mexico-America, India, while some refer to smaller territories, Switzerland, Cyprus, Antioch, Egypt.

We have devoted a whole chapter to the problems inherent in inclined dials. The interest of multi-faced dials lies in the application of the solutions to these problems and this study can obviously lead to many agreeable hours spent in research. Plate 41 shows the unfolding (done by the author) of the whole of the dial of Mont Sainte-Odile in so far, at least, as the weathering of the stone and the moss which partially covers it allowed; the job could not be done with any great accuracy. The dial has been in its present location only since 1935. Originally it was in the Cistercian abbey of Neubourg near Haguenau (Bas-Rhin), where it was built during the first half of the eighteenth century; later it was for many years in the garden of a seminary near Strasbourg.

The same city of Haguenau has a similar dial, perhaps of the same origin but not as fancy, in the Saint George cemetery. Still another, equipped with scaphions (p. 10) but very old and with its styles missing has found a temporary refuge in the Rohan palace museum in Strasbourg.

It is fitting to mention in this chapter the famous analemmatic dial on the parvis of the church in Brou as well as its counterpart in Dijon (already discussed at length in chapter six). As mentioned previously, this type of dial was useful mainly because its presence on the table of a portable horizontal dial eliminated the need of a compass. As a dial for the use of the public, it was rather a curiosity, because especially during daytime when the crowd was large it must have been difficult to read, in any event more difficult than a wall dial.

Our task would be incomplete if we did not also mention some gigantic dials in masonry which, in conjunction with other instruments of the same nature, make up the peculiar observatory built by the rajah Jaih Singh II around 1787 in Jaipur, a city in India close to Delhi. The author is sorry not to have been able to obtain from the authorities of that country photographs giving details of this most interesting installation. There are equatorial dials for observations of various kinds and the one which interests us most is an equatorial dial very much like the one shown in figure 105. It consists of a style in the shape of a triangle but of sufficient thickness to carry on its crest convenient steps which allow one to climb to a small pavilion on the summit. The style is more than

100 feet (30 metres) long with marble ridges. The equatorial circle, also marble, on the inside and many metres in diameter, is divided into sectors with their centres on the respective ridges of the style and naturally the axis of the cylinders making up these sectors is parallel to the style and to the axis of the world.

This sundial, as far as we know, is by far the largest ever built. At the beginning of this century it and the whole of the observatory were in ruins. It was restored around 1901 by the British administration and now seems to be in fair condition.

FIGURE 105 The great solar dial of the Observatory of Jaipur

Jaih Singh II must have been a rather extraordinary man for his times and country: he had four other observatories built, of less importance than the one in Jaipur, it is true, in Delhi, Benares, Mathura, and Ujjain. When some Portuguese missionaries spoke to him about recent developments in astronomy in Europe, he got in touch with King Emmanuel and the two sovereigns exchanged the results obtained by their scientists. The tables compiled in the observatories of the rajah turned out to be more accurate than those of de la Hire; this was attributed, probably with reason, to the smaller dimensions of the instruments used in Europe.

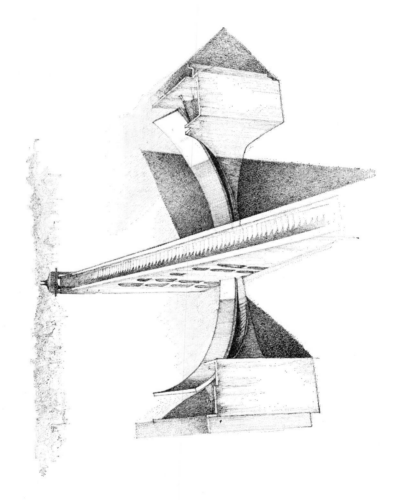

Plate 5 reproduces a beautiful example of a French dial dating probably from the fifteenth or perhaps sixteenth century; it is found on an old wall in the court of the city hall of La Rochelle. Owing to the height of the surrounding constructions, this dial, rather well preserved in spite of the weather and the poor quality of the stone, can only indicate the time from 10 o'clock in the morning until 3 o'clock in the afternoon. We note that it carries lines for the quarter hours. The hole at the tip of the style seems superfluous since there are no daily arcs; nevertheless it may have been useful because of the monotone colour of the stone, for it would prevent the shadow of the style becoming indistinguishable from one of the numerous hour-lines or becoming invisible in the rather dense network of lines.

When we were discussing the two multi-faced dials from the district of Haguenau, we saw that there is a tendency to assign to them a common origin. Indeed, the passage of an active and ingenious gnomonist through a particular region often resulted in a whole swarm of dials marked by the personal touches of this creator. So, in very many parts of the country not a single dial is to be found. In some villages, however, there are three or four dials at a time in the immediate vicinity. Sometimes, too, some gnomonist was encouraged by local conditions, as was Peter Anich, a peasant from Tyrol who lived between 1723 and 1766 and who, finding interest in mathematics, became a geometer and surveyor and spent most of his time making sundials.

He had the good fortune to live in a country where people loved music and singing and where they usually decorated their houses with coloured paintings and various inscriptions. We may say that Peter Anich took advantage of this fact far beyond what would be expected from a single worker. A great number of his dials are still almost intact, and they may certainly be counted among the most beautiful realizations of the popular arts to which

gnomonics has given rise, even taking account of the special and lively folklore of the Tyrolean population and of its religious character. Naturally Peter Anich was not the only one engaged in this trade. Professor Kühnelt of the University of Innsbruck has studied the problem of conserving this part of his country's heritage and he has listed in one of his numerous publications 176 ancient dials in northern Tyrol alone that are worthy of attention.

Plates 13 to 16 and Plate 20 give an idea of these Tyrolean dials whose original character and vivid colours attract the attention of even the most indifferent tourist.

On the dial of the church of Telfes represented in Plate 13, strips placed in the middle of the group of hyperbolas indicate the length of the day and the elevation of the sun at noon. To the right and to the left, between the daily arcs, the signs of the zodiac are shown with their symbols. The motto can be translated as "You know not the day nor the hour." The dial, although unsigned, is usually attributed to Anich. The baroque style of its frame is found again in the dial of the church of Natters, shown in Plate 16; it has the same author and the two centuries elapsed have not dulled the strength of the colours. As for the previous dial, it gives on the central band the length of the day, but not the elevation of the sun. The motto originally entered on the frame would have referred to the painting which shows the Creator floating over the water. It was replaced later by the one shown in our plate which reads, "Use time as wise men and not like fools because the days are evil" (Eph. 5:16).

Plate 20 shows another dial drawn by Anich which is found on an inn at Gschnitz and which has been slightly damaged, as the photograph indicates, by the dampness of the masonry of the wall. It is an ordinary dial whose hour-lines are in the frame and are barely visible. Between the hyperbolic lines of the solstices, Anich has traced the Babylonic and Italic lines. The numbering of the first set goes from 12 – the 0 Babylonic hour – to 8; the hour-line 12 (or 0) is horizontal (see p. 92). The Italic lines go from 11 to 22. According to Dr Kühnelt, this dial is the only one of its type in northern Tyrol.

Another rather singular dial is the one at Prutz shown in Plate 14; it too is drawn on the wall of an inn and it has the peculiarity of showing a division of the day into temporary hours in addition to the normal dial and an astrological table like that in figure 91. Chapter five deals at length with this extremely rare type of dial. The dial at Prutz is certainly one of the curiosities of gnomonics: its vivid and varied colours make it highly picturesque and its complexity is intriguing.

Plate 15 shows the very beautiful dial of the castle of Ambras near Innsbruck. It declines slightly to the northwest so that its nearly parallel lines converge towards the centre which is located to the right and below, outside the dial. The appearance of this dial is quite unique.

The famous dial of Queens College in Cambridge which dates in its actual form to around 1733 has been mentioned several times. The moist climate of England has forced its owners to restore it often, but they have always preserved the original tracing and colours. This, at least, is maintained by Dr G. C. Shephard of Birmingham who has studied it and kindly helped us to describe it in detail.

From many points of view, this dial is a beautiful achievement in the field of gnomonics. As seen in Plate 9, it appears overloaded, but, in fact, a judicious choice of colours makes it very easy to read as well as extremely attractive. Its exterior frame carries – in gold on a blue background – the Roman numerals of the true hours. The equation of time is not mentioned for the good reason that in 1733 the concept of mean time was known and used only by astronomers. Its introduction into daily life came only later.

We have already mentioned in chapter seven the use of this dial for reading time from the shadow of the moonlight. The following is an example of this reading quoted from Dr G. C. Shephard's very interesting booklet *Queens College Dial*:

For example, on February 26th, 1948, at about 11 PM the moon was casting a shadow that gave the reading 9.35 on the outer scale. There was a full moon on Tuesday, 24th February at 5.16 PM so that the moon was about $15 + 2\frac{1}{4}$ days old. Its hour angle when 17 days old (from the table) was 1.36, and allowing 12 minutes for the quarter day, the moon's hour angle at the time of observation was 1.48. The equation of time on February 26th is +13 minutes, so that the GMT was $9.35 + 1.48 + 0.13 = 11.36$ PM. In fact the clock showed 11.50. This may not seem very accurate, but if you arrive at the time correct within half an hour, you may be proud! The moon's motion is so irregular that no moon-dial can possibly be accurate, and is to be regarded as providing an exercise in mental arithmetic rather than being an instrument of any practical value.

The reader will recall that our conclusions in chapter seven were entirely similar.

The author(s?) of this dial seemed to have intended to make it a very general gnomonical instrument: the vertical lines of the solar azimuth have been traced and numbered not in degrees but in compass directions, using the English abbreviations, of course. These vertical lines are drawn in black. They are completed by a group of hyperbolas drawn in red and carrying in degrees the signs for the elevation of the sun. The shadow of a little wooden ball borne by the style is used for the reading of these two quantities, as well as for all the others with the exception of the hours.

We only mention the hyperbolas of the signs of the zodiac which this dial has in common with very many others. But let us mention here that the signs to be used during an observation are those on the right when the days are getting longer and those on the left when they are becoming shorter. The same remark applies to the narrow column on the edge of the dial which is reserved for the months of the year.

Another original feature of the dial of Queens College is the column ORTUS SOLIS (sunrise) where the hours of sunrise are indicated beside the arcs of the signs of the zodiac and the column LONGITUDO which gives beside the same arcs the right ascension of the sun. Of course, it is expressed in hours. Here, unfortunately, as noted by Dr Shephard, an error has been made: the angles are measured, not from the vernal point, but rather from the fall equinox, the point diametrically opposed to it on the ecliptic. A correction must then be made for each reading. Indeed it is necessary to (1) add 12 hours to the readings made from December 21 to March 21, (2) subtract 12 hours from March 21 to June 21, and (3) subtract the reading from 24 hours from June 21 to December 21.

Finally, a group of hour-lines indicates the temporary hours, as seen on the Tyrolean dial at Prutz. Let us repeat again that dials carrying these lines are extremely rare. Again as at Prutz, a table of the planets, masters of the hours of the day, is sketchily indicated by the planetary signs beside the signs of the zodiac. This table, as is to be expected, is quite incomplete.

We have remarked that the authors of this dial wanted to be thorough. Nevertheless, they have omitted the Babylonic and Italic lines with their valuable information on sunrise and sunset and the length of the day.

While driving through France, the tourist may sometimes be surprised by coming across an old dial perched on a column, right in the middle of a modern intersection, which goes on giving, as it always has, the true time to the passers-by. Such a witness of the old days (Plate 42) may be found in Alsace near Dorlisheim: from the cross-roads that it dominates, it has seen

Cossacks go by in 1813, Prussians in 1870, and American tanks in 1944.

Under more clement skies, above Nice and its blue sea, an original dial dug out of a white marble plate 2.45m wide, is found in the convent of Cimiez (Plate 19). The dial bears the date of its construction, 1876, and the name of its builder Father Ildéphonse Récollet. It is only one of the dials to be found in this convent, but it is the most interesting because Father Ildéphonse traced the meridians of mean time on each of the hour lines with great care. He made the reading of the date easier with the help of a scale at five-day intervals on the extreme right and extreme left edges of the dial. Because of its large dimensions, the results of the readings are astonishingly accurate. We note the strange deformation of the meridians and also the fact that Father Ildéphonse reserved the whole of the dial lying between the solstices for the indication of the mean time; he reduced the reading of the true time to that part of the dial located between the centre and the winter solstice.

The dial of Father Ildéphonse is about the only one of its kind. Among the French dials it is indeed a beautiful piece for study and teaching. However, for the mass of information which a dial can supply, it lags far behind the veritable monument which Father Bonfa, another monk, conceived and made in 1673 in a Jesuit college at Grenoble, which is now a high school for girls. Probably this dial has no counterpart in the world either for its originality or for the diversity of the information which it can supply to those who care to study it.

Father Bonfa was born in Nimes in 1638. We do not know why he replaced the dial by a small mirror and the shadow by a ray of light. Might it be that a glass of water was forgotten on the window sill of his cell? The glass would have reflected the sun's rays inside the cell and he would have noticed the light moving on the wall. In any event, he put a small mirror on the window of a staircase and studied the path of the light which it reflected. Finally, he traced in many colours a complicated system of lines and of abacuses in all the places which the ray could penetrate, around and even under the climbing staircase, the description of which could easily fill a small book. Recalling that the precision of the reading is proportional to the length of the reflected ray, we see that therein lies the genius of Father Bonfa.

This unique dial covers more than 100 square metres and it indicates in its abacuses and lines the following information: true local time, Babylonic time, Italic time, times of sunrise and sunset, the beginning of dawn and the end of dusk, the months of the year and the signs of the zodiac, the approximate date, the position of the 12 astrological houses, a lunar calendar with a table of the epact and a civil calendar giving the date of the lunar day, an "Horologium novum" showing the position of the moon for any hour for a certain number of points on the earth, and finally a calendar of the victories of Louis XIV, one of the feasts of the Holy Virgin, a table of the Jesuits who had been canonized, and a graph giving the simultaneous time in various Jesuit colleges over the whole world.

It is evident that Father Bonfa was limited only by space, for of time he had plenty. Naturally, much of the information on his monumental work contained in the above list has nothing to do with gnomonics. But the idea was new and the author took maximum advantage of it. It is not surprising that it is difficult to find a way through the markings of this gigantic dial, since the basis was new and complicated. Its description would take us too far afield and its photographic reproduction is made difficult by the shape of the area used. In 1921, the *Bulletin of the Scientific Society of Isère*, Volume 42, published a paper by J. Ray-Pailhade, A. Rome, and Aug. Favot, giving very interesting details about it

as well as colour plates. Their examination helps us to get some idea of the work of Father Bonfa and to understand it better. This publication is recommended to the interested reader.

There is scarcely a city in France where the gnomonics enthusiast does not end up by discovering at least one old dial on an ancient building. Very often it will carry some motto reflecting the philosophy of its constructor. We show one example among thousands in Plate 12. This dial is found in the courtyard of the old abbey of Cluny. Extremely simple, meant at most to indicate the time, it does not fail, however, to warn visitors, who have replaced the Benedictine monks of past days, of the hour that will be their last: *Ex iis unam cave*, meaning *Beware of one of them!*

Sometimes these dials, such as the one that decorates a promenade on the shore of the lake at Annecy and whose curious diversity attracts the attention of passers-by, are more recent. It is the work of Jean-Marie Dumurger, a Capucin called Brother Arsène. He was born in Annecy in 1808 and died there in 1876. While living in the religious order, he practised the trade of stone sculptor. The number of remarkable dials made by clerics is evidence of a relationship between the search for meditation and the practice of gnomonics.

In his Annecy monument, made of local stone, Brother Arsène installed a profusion of dials. At the tip, a thick star with seven branches is held by a support placed on a broader base. The two planes which make up the faces of the star are parallel to the plane of the equator so that its upper face can carry an equatorial dial for use in summer. For the winter, when the declination of the sun is negative, two other equatorial dials have been designed on the inferior face of the two upper branches of the star (Plates 45 and 46). The later planes of the branches of the star have been designed to serve as tables for polar dials for which the external edges act as styles. These edges and the tables are indeed perpendicular to the equator and therefore parallel to the axis of the world. The support on the south of the star carries a meridional dial, on the north a septentrional dial; the two side faces are set with an oriental and an occidental dial, respectively.

The south face of the large base has a dial basically meant to yield information of a particular nature. A meridian occupies its centre; it is placed on the vertical line of noon of Annecy and is crossed by daily arcs labelled with the signs of the zodiac. Finally, the edge of the table carries, between the digits of the hours, a certain number of marks whose location fixes the hour-lines of the instants when it is local noon in the main capitals of Europe.

Erected in 1874, "Brother Arsène's gnomon," as it is known in rather Annecy, cannot be considered any ancient piece (in spite of its rather Spartan inscriptions). But the general shape of its base, displaying the ornate style of the buildings of the late nineteenth and early twentieth century, is certainly *passé*, especially when we consider the magnificent dial recently installed in the Botanical Gardens at Bremen. This is a large but simply horizontal dial, cut in a piece of granite and carrying the twelve signs of the zodiac (Plates 48 to 50). The simple and clear way in which the artist explains the correction of the equation of time to be applied to the hour read in order to obtain the mean time is noteworthy. Indeed, in Plate 48, we see the detail of a graph whose abscissa indicates the months of January to December and whose ordinate indicates the minutes. The inscription reads: "The dial (literally: the watch) is slow ..." then under the abscissa in the negative portion: "is fast." The whole of the graph and the inscription make us understand that the correction must be added if the equation of time is positive and subtracted if it is negative.

Two other modern dials in moulded and coloured concrete are shown in Plates 10 and 11. In both cases, the Babylonic and Italic lines have been traced, so that the time of sunrise and of sunset and

the length of the day can be read off in the way indicated previously. The concrete is reinforced, the thickness of the plate is 45 mm, and the dimensions are approximately 1×0.75 m. The vertical dial found on the façade of a building in Carcassonne has been set with the graph described in chapter seven for use in moonlight.

Although this type of dial, which provides multiple information, may interest ardent gnomonists, it must be admitted that the recent trend on public buildings is to dials which generally give only the time. Such a dial reduced to the bare minimum may have its place in a small town or village church (Plate 22). If it is to be put on the wall of a school, as is the case in Plate 21, it is certainly wrong to embellish it with decorations that have no gnomonic bearing whatsoever; the coat of arms of the city, if it makes any sense, could be used. It seems clear that from an instructional point of view dials such as those shown in Plates 10 and 11 would be preferable.

There is a tendency now, as indicated by some very modern creations, to return to the technically simpler applications of the equatorial dial. The greater precision of the readings is the pretext for its use. But, sometimes, the gnomonic essentials are relegated to the background and the real intention is to create a work of art of appreciable aesthetic and symbolic value, for instance the magnificent armillary sphere found in the park of Phillips Academy in Andover, Massachusetts (Plate 51). Here is a description drawn from information kindly supplied by the secretary of Academy, Mrs Anny R. Robinson:

The Armillary Sphere is the symbol of the world. It traces in its bands a vertical Meridian supporting the equatorial belt with its diagonal Ecliptic and the Arctic and Antarctic Circles. The centers of these latter are the poles themselves. The Sphere is placed on its base at the angle of the earth – 42 degrees, 30 minutes – at Andover, with the axis pointing due north and south. The star at the extremity of the north points to-

Remarkable dials | 125

wards and symbolizes the North Star. The shaft of the axis of the world is the gnomon, whose shadow on the belt of the equator indicates the time of the day, by the position of the sun, so when the sun is due south, the hour is noon. Opposite to noon is midnight, which is represented on the inner side of the band by the moon, on the one side of which is Evening, with her owl, and on the other Dawn and the cock. At the halfway points between noon and midnight, are Six o'clock of the Morning and Six o' clock of the Evening. As in the celestial sphere the path of the sun is shown by the ecliptic with the signs of the zodiac, each corresponding to a month of the year, so here the sun's progress among the stars is symbolized and the signs of the zodiac are shown in the figures in high relief on the band of the equator, with their names on the ecliptic. The four Elements figure in the scheme; Water in the wave motif on the base, with the Earth motif growing out of it, while Air is represented by the ribbon motif on the circles, and Fire in the flames encircling in the whole on the vertical Meridian. The group of Man, Woman, and Child in the center symbolizes the Cycle of Life as the Sphere itself is the Cycle of Eternity – it is Humanity within the Universe, or Humanity and Eternity.

In the Scandinavian countries, one sees in the gardens and the public and private parks small pieces representing armillary spheres, whose style, i.e., the axis of the world, is, however, very often removed. The gnomonic value of the apparatus is thus forgotten and only its decorative purpose is retained. The reproduction of a very complete armillary sphere of the seventeenth century is given in Plate 38 as it appeared in the celebrated book of Dom Bedos de Celle. None of the circles concerned with celestial motion have been omitted: the axis of the world is represented with a little ball standing for the earth on which the shadow of the tropics falls during the solstices and that of the equator during the equinoxes.

In conclusion we shall mention a very modern instrument which recalls the heliochronometers of the French railroads at the end of the nineteenth century; a fine sample may be found in

Lucerne by the lake on the Schwanenplatz. Plate 47 does not represent this piece exactly, but it shows its gnomonic features rather well.

This is a very refined equatorial dial whose table carries the hour-lines at intervals of 20 minutes. The cylindrical table, whose inferior and superior edges represent the tropics of an imaginary armillary sphere, has been used to hold the ecliptic, the pictures of the signs of the zodiac, and the graph of the equation of time. For the former, the scale is in degrees; for the latter, in minutes; the abscissae of the hour-lines of the 20-minute intervals separate periods of 10 days.

We note, with a magnifying glass if need be, that the instrument is set for the normal true time, i.e., for the local true time corrected for the longitude of Lucerne to obtain the legal time. The observer has only to add or subtract the equation of time which may be easily read on the graph, if he knows the date.

The style, in the shape of an arrow directed towards the pole of the sky, carries in the centre of the sphere a disc pierced with a hole whose very sharp luminous trace on the table of the dial allows relatively accurate readings. In addition, the style may be rotated on its bearings (at the poles) so that the rays of the sun always fall almost perpendicularly on the dial, as shown by the circular reflection of the hole.

This instrument, recalling the motion of the sun, is instructive and useful, since it supplies the legal time with a close precision to the minute; others like it have been adopted by many cities which have set them up in frequented areas. Since they are made entirely of rustproof metallic alloys, they keep their shining newness. They are increasingly popular as a kind of living sculpture or public ornament. The public dial of Lucerne, it is interesting to note, is accessible from any side and can be touched by all, but despite exposure day and night for years no trace of deterioration has ever been noted.

CHAPTER TEN

Inscriptions on Sundials

Forty centuries of civilization have brought the sundial to its present state of perfection. For forty centuries people have looked at sundials. The sundial has witnessed the birth and disappearance of the clepsydra and the sand-clock, of the fire and the oil clocks, and it has survived them all.

Indeed, in spite of the high precision of mechanical watches, chronometers, and clocks, the man of today continues to look at the sundial. He looks at it with increased interest, even with respect and some awe; no doubt because he has inherited from the long past some of the thoughtfulness to which the contemplation of its eternal motion seems always to lead.

A sundial is a living object. It needs no winding and is driven by no weight. It has something to say and says it. It speaks about time, never ceasing to recall the flight of time, its tragedy and irreversibility for man. The thoughts arise of death, of the end of everything, of eternity, of the world beyond.

These thoughts were expressed early in thousands of inscriptions which have appeared on dials. They vary according to the philosophy of the builder or the patron; they allude to philosophical or religious questions, to the earth and the sun, to the day and the night; sometimes, but rarely, they take a hedonistic twist. Most of them are written in Latin, the language of serious people,

of scientists, and of clerics.

These inscriptions, called sundial quotations, tend to disappear little by little as buildings and their dials get older, regrettable as it is. They do indeed constitute a bit of folklore, a popular philosophy, the remnant of a primitive literature with its roots deep in the hearts of our ancestors.

This has been understood, among others, by Charles Boursier who compiled a book, *800 Devises de cadrans solaires* (Paris, 1936), in which he collected with devotion and infinite patience all he could find in Europe of the fragments of this historical treasure. His little book has been out of print for a long time and demands a high price among collectors, though it is available in many public libraries in France.

Space does not allow us to give too many of these quotations here, but it seems to us that this book would not be complete if it did not contain at least some of them. We also hope that one of these quotations will be used in the proper surroundings on each dial that this book will help to bring into existence.

In keeping with tradition, the quotations are given in the original Latin; the English translation which we give can be considered at most as a motto. The choice will be governed by taste and surroundings; the text is usually shorter in Latin.

CAUTE CAVE MEDIO NE DESIT LUMINE LUMEN	Be careful that the light does not fail you during the day
SOLE ORIENTE FUGIUNT TENEBRAE	The shadows flee when the sun approaches
CUI DOMUS HUIC HORA	Each household should have the time
SI SOL DEFICIT, RESPICIT ME NEMO	Nobody looks at me when the sun is not there
MEAM VIDE UMBRAM, TUAM VIDEBIS VITAM	Look at my shadow, you will see your life
SOL IMMOBILIS TERRA VERSAT	The motion of the earth makes the sun rotate
ULTIMA LATET UT OBSERVENTUR OMNES	The last is hidden so that we have to watch them all
FESTINA LENTE	Make haste slowly
VIDI NIHIL PERMANERE SUB SOLE	I have seen nothing last forever under the sun
POST LABOREM REQUIES	Rest after work
ALTERA PARS OTIO, PARS ISTA LABORI	Devote the other to leisure, devote this one to work
HORA FUGIT NE TARDES	Time flies, do not be late
UTERE NON REDITURA	Use it, it won't come back
SERIUS EST QUAM COGITAS	It is later than you think
SIC VITA FLUIT, DUM STARE VIDETUR	Life goes and still it always looks the same
TEMPUS EDAX RERUM	Oh time, devourer of things
TEMPUS VOLAT, HORA FUGIT	Time flies, the hour escapes
TEMPUS BREVE EST	Time is short
SIT FAUSTA QUAE LABITUR	Let this hour be favourable
VIVERE MEMENTO	Do not forget to live
DONA PRAESENTIS RAPE LAETUS HORAE	Take quickly the gifts of the present hour
UTERE, NON NUMERA	Use them, don't count them
POST TENEBRAS SPERO LUCEM	Oh light, I hope for thee in this darkness
DUM TEMPUS HABEMUS OPEREMUR BONUM	Let us do some good while there is still time
OMNES AEQUALES SOLA VIRTUTE DISCREPANTES	They are all the same, only your good deeds make them different
VER NON SEMPER VIRET	Spring does not last forever
UMBRA SICUT HOMINIS VITA	The life of man is like a shadow
OMNES VULNERANT, ULTIMA NECAT	Each one wounds, the last one kills
SOL OMNIBUS LUCET	The sun shines for all
RUIT HORA	Time flies
EX IIS UNAM CAVE	Beware of one of them
MEMOR ESTO BREVIS AEVI	Remember that life is short

LUCEM DEMONSTRAT UMBRA — It is light that makes a shadow

AMICIS QUALIBET HORA — Any hour can go to my friends

TEMPUS OMNIA DABIT — Time will eventually give everything

UNA EX HIS ERIT TIBI ULTIMA — One of these will be your last

TEMPUS VINCIT OMNIA — Time conquers all

VITA IN MOTU — Life is in motion

SIC LABITUR AETAS — So goes life

LENTE HORA, CELERITER ANNI — The hour is slow but the years run quickly

TEMPUS FUGIT — Time flies

FRUERE HORA — Enjoy this moment

UNA DABIT QUOD NEGAT ALTERA — One will give what the other has refused

HORAS NON NUMERO NISI SERENAS — I show only the bright hours

VITA SIMILIS UMBRAE — Life is like the shadows

DOCEO HORAS — I tell the time

SOL ME PROBAT UNUS — Only the sun can prove that I am useful

SOL REX REGUM — O sun, king of kings

Boursier quotes an English inscription which we reproduce here for its poetry and optimism:

Let others tell of storms and showers
I only mark the sunny hours

It has its counterpart in German where it is used as a proverb rather than as a quotation on a sundial:

MACH' ES WIE DIE SONNENUHR
ZÄHL' DIE HEITEREN STUNDEN NUR

Do like the sundial:
count only the bright hours

Another quotation attracts our attention – and good humour – by its amusing, precise, and imperative meaning:

ARRESTO TI, PASSANT, REGARDO QUANTO ES D'OURO ET ...
FOUTO ME LOU CAMP

Passer-by, stop, look at the time, and ...
get the hell out of here!

Only those who do not like Provence will fail to understand this.

In conclusion, let us recall a final quotation, very well known and perhaps the most used of all, which the author himself has placed on some dials:

CARPE DIEM   *Use well the day*

We had a reason for keeping it until the end because it gives us a chance to illustrate the fact that Latin quotations may give rise to highly unexpected interpretations.

Indeed at the beginning of the last century, this quotation figured on the sundial of the house of an old eccentric, located in a twisting, rather long, thinly populated street of a suburb of Strasbourg, well-known now for the extent of its gardens and the quality of its vegetables. Time and again, the gardeners passing by on their carts have looked at the mysterious inscription, wondering about its meaning. In vain, it seems ... A century has gone by and the house with its dial have long since disappeared. But the street still exists. It is called *rue de la Carpe* ...

TABLE 1

Conversion of degrees and minutes of arc into units of time

MINUTES OF ARC	SECONDS AND MINUTES (TIME)	DEGREES OF ARC	HOURS AND MINUTES (TIME)
1'	4 s	1°	4 m
2'	8 s	2°	8 m
3'	12 s	3°	12 m
4'	16 s	4°	16 m
5'	20 s	5°	20 m
6'	24 s	6°	24 m
7'	28 s	7°	28 m
8'	32 s	8°	32 m
9'	36 s	9°	36 m
10'	40 s	10°	40 m
20'	1 m 20 s	15°	1 h
30'	2 m 00 s	20°	1 h 20 m
40'	2 m 40 s	25°	1 h 40 m
50'	3 m 20 s	30°	2 h
60'	4 m 00 s	35°	2 h 20 m
		40°	2 h 40 m
		45°	3 h
		50°	3 h 20 m
		55°	3 h 40 m
		60°	4 h
		65°	4 h 20 m
		70°	4 h 40 m
		75°	5 h
		80°	5 h 20 m
		85°	5 h 40 m
		90°	6 h

TABLE 2

Atmospheric refraction

OBSERVED ELEVATION	REFRACTION (TO BE SUBTRACTED)
5°	10' 2
10°	5' 5
15°	3' 7
20°	2' 7
25°	2' 1
30°	1' 7
35°	1' 4
40°	1' 2
45°	1' 0
50°	0' 8
55°	0' 7
60°	0' 6
65°	0' 5
70°	0' 4
75°	0' 3
80°	0' 2
85°	0' 1
90°	0' 0

TABLE 3

Mean value of the equation of time, in minutes (at true noon)

DAY	JAN.	FEB.	MAR.	APR.	MAY	JUNE	JULY	AUG.	SEPT.	OCT.	NOV.	DEC.
1	+3.4	+13.6	+12.5	+4.1	−2.8	−2.3	+3.6	+6.3	+0.2	−10.1	−16.3	−11.2
2	3.9	13.7	12.3	3.8	3.0	2.2	3.8	6.2	−0.1	10.4	16.4	10.8
3	4.3	13.8	12.1	3.5	3.1	2.0	4.0	6.2	0.5	10.8	16.4	10.4
4	4.8	13.9	11.9	3.2	3.2	1.9	4.2	6.1	0.7	11.1	16.4	10.0
5	5.2	14.0	11.7	2.9	3.3	1.7	4.4	6.0	1.1	11.4	16.4	9.6
6	5.7	14.1	11.5	2.6	3.4	1.5	4.6	5.9	1.5	11.7	16.3	9.2
7	+6.1	+14.2	+11.2	+2.3	−3.4	−1.3	+4.7	+5.8	−1.8	−12.0	−16.3	−8.8
8	6.5	14.2	11.0	2.1	3.5	1.2	4.9	5.7	2.1	12.3	16.3	8.3
9	6.9	14.3	10.7	1.8	3.6	1.0	5.0	5.5	2.5	12.6	16.2	7.9
10	7.3	14.3	10.5	1.5	3.6	0.8	5.2	5.4	2.8	12.8	16.1	7.5
11	7.8	14.3	10.2	1.2	3.7	0.6	5.3	5.2	3.2	13.1	16.0	7.0
12	8.2	14.3	10.0	0.9	3.7	0.4	5.4	5.1	3.5	13.4	15.9	6.5
13	+8.5	+14.3	+9.7	+0.7	−3.7	−0.2	+5.6	+4.9	−3.9	−13.6	−15.8	−6.1
14	8.9	14.3	9.4	0.4	3.7	0.0	5.7	4.7	4.2	13.8	15.6	5.6
15	9.3	14.2	9.1	+0.2	3.7	+0.2	5.8	4.5	4.6	14.1	15.5	5.1
16	9.6	14.2	8.9	−0.1	3.7	0.4	5.9	4.3	5.0	14.3	15.3	4.6
17	9.9	14.1	8.6	0.2	3.7	0.7	6.0	4.1	5.3	14.5	15.1	4.1
18	10.3	14.0	8.3	0.5	3.7	0.9	6.1	3.9	5.5	14.7	14.9	3.6
19	+10.6	+13.9	+8.0	−0.7	−3.6	+1.1	+6.2	+3.7	−6.0	−14.9	−14.7	−3.2
20	10.9	13.8	7.7	0.9	3.6	1.3	6.2	3.5	6.4	15.1	14.5	2.7
21	11.2	13.7	7.4	1.2	3.5	1.5	6.3	3.2	6.7	15.2	14.3	2.2
22	11.5	13.6	7.1	1.4	3.5	1.7	6.3	3.0	7.1	15.4	14.0	1.7
23	11.8	13.5	6.8	1.6	3.4	2.0	6.4	2.8	7.4	15.6	13.7	1.2
24	12.0	13.4	6.5	1.8	3.3	2.2	6.4	2.5	7.8	15.7	13.4	0.7
25	+12.3	+13.2	+6.2	−1.9	−3.2	+2.4	+6.4	+2.2	−8.1	−15.8	−13.1	−0.2
26	12.5	13.1	5.9	2.1	3.1	2.6	6.4	1.9	8.4	15.9	12.9	+0.3
27	12.7	12.9	5.6	2.3	3.0	2.8	6.4	1.7	8.8	16.0	12.5	0.8
28	12.9	12.7	5.3	2.4	2.9	3.0	6.4	1.4	9.1	16.1	12.2	1.3
29	13.1		5.0	2.6	2.8	3.2	6.4	1.1	9.5	16.2	11.9	1.8
30	13.3		4.7	2.7	2.6	3.4	6.4	0.8	9.8	16.3	11.5	2.3
31	+13.4		+4.4		−2.5		+6.3	+0.5		−16.3		2.8

After the *Ephémérides Nautiques* for the year 1963.

TABLE 4

Mean value of the solar declination (for 1962, noon GMT)

DAY	JAN.	FEB.	MAR.	APR.	MAY	JUNE	JULY	AUG.	SEPT.	OCT.	NOV.	DEC.
1	−23°01′	−17°09′	−7°40′	+4°28′	+15°01′	+22°02′	+23°08′	+18°04′	+8°22′	−3°06′	−14°22′	−21°46′
2	−22°56′	−16°52′	−7°17′	+4°51′	+15°19′	+22°10′	+23°03′	+17°49′	+8°00′	−3°30′	−14°41′	−21°56′
3	−22°50′	−16°35′	−6°54′	+5°14′	+15°37′	+22°17′	+22°59′	+17°34′	+7°38′	−3°53′	−15°00′	−22°04′
4	−22°44′	−16°17′	−6°31′	+5°37′	+15°54′	+22°25′	+22°54′	+17°18′	+7°16′	−4°16′	−15°19′	−22°13′
5	−22°38′	−15°59′	−6°08′	+6°00′	+16°12′	+22°31′	+22°49′	+17°02′	+6°54′	−4°39′	−15°37′	−22°21′
6	−22°31′	−15°40′	−5°44′	+6°22′	+16°29′	+22°38′	+22°43′	+16°46′	+6°31′	−5°02′	−15°55′	−22°28′
7	−22°24′	−15°22′	−5°21′	+6°45′	+16°45′	+22°44′	+22°37′	+16°29′	+6°09′	−5°25′	−16°13′	−22°35′
8	−22°16′	−15°03′	−4°58′	+7°08′	+17°02′	+22°50′	+22°30′	+16°12′	+5°46′	−5°48′	−16°31′	−22°42′
9	−22°08′	−14°44′	−4°34′	+7°30′	+17°18′	+22°55′	+22°23′	+15°55′	+5°24′	−6°11′	−16°48′	−22°48′
10	−21°59′	−14°25′	−4°11′	+7°52′	+17°34′	+23°00′	+22°16′	+15°38′	+5°01′	−6°34′	−17°05′	−22°54′
11	−21°50′	−14°05′	−3°47′	+8°14′	+17°50′	+23°04′	+22°08′	+15°20′	+4°38′	−6°57′	−17°22′	−22°59′
12	−21°40′	−13°45′	−3°24′	+8°36′	+18°05′	+23°08′	+22°00′	+15°02′	+4°16′	−7°20′	−17°39′	−23°04′
13	−21°31′	−13°25′	−3°00′	+8°58′	+18°20′	+23°12′	+21°52′	+14°44′	+3°53′	−7°42′	−17°55′	−23°08′
14	−21°20′	−13°05′	−2°37′	+9°20′	+18°35′	+23°15′	+21°43′	+14°26′	+3°30′	−8°04′	−18°11′	−23°12′
15	−21°09′	−12°44′	−2°13′	+9°42′	+18°49′	+23°18′	+21°34′	+14°07′	+3°07′	−8°26′	−18°26′	−23°16′
16	−20°58′	−12°24′	−1°49′	+10°03′	+19°03′	+23°21′	+21°24′	+13°49′	+2°43′	−8°49′	−18°41′	−23°19′
17	−20°47′	−12°03′	−1°25′	+10°24′	+19°17′	+23°23′	+21°14′	+13°29′	+2°20′	−9°11′	−18°56′	−23°21′
18	−20°35′	−11°42′	−1°02′	+10°45′	+19°30′	+23°24′	+21°04′	+13°10′	+1°57′	−9°33′	−19°11′	−23°23′
19	−20°22′	−11°21′	−0°38′	+11°06′	+19°43′	+23°25′	+20°53′	+12°51′	+1°34′	−9°54′	−19°25′	−23°25′
20	−20°09′	−10°59′	−0°14′	+11°27′	+19°56′	+23°26′	+20°42′	+12°31′	+1°11′	−10°16′	−19°39′	−23°26′
21	−19°57′	−10°38′	+0°09′	+11°47′	+20°09′	+23°27′	+20°31′	+12°11′	+0°47′	−10°37′	−19°52′	−23°26′
22	−19°43′	−10°16′	+0°33′	+12°08′	+20°21′	+23°26′	+20°19′	+11°51′	+0°24′	−10°59′	−20°05′	−23°27′
23	−19°29′	−9°54′	+0°57′	+12°28′	+20°32′	+23°26′	+20°07′	+11°31′	+0°01′	−11°20′	−20°18′	−23°26′
24	−19°15′	−9°32′	+1°20′	+12°48′	+20°44′	+23°25′	+19°55′	+11°11′	−0°23′	−11°41′	−20°31′	−23°25′
25	−19°00′	−9°10′	+1°44′	+13°07′	+20°55′	+23°24′	+19°42′	+10°50′	−0°46′	−12°02′	−20°43′	−23°24′
26	−18°46′	−8°47′	+2°08′	+13°27′	+21°05′	+23°22′	+19°29′	+10°29′	−1°10′	−12°22′	−20°54′	−23°22′
27	−18°30′	−8°25′	+2°31′	+13°46′	+21°16′	+23°20′	+19°16′	+10°08′	−1°33′	−12°43′	−21°05′	−23°20′
28	−18°15′	−8°02′	+2°55′	+14°05′	+21°26′	+23°18′	+19°02′	+9°47′	−1°56′	−13°03′	−21°16′	−23°18′
29	−17°59′		+3°18′	+14°24′	+21°35′	+23°15′	+18°48′	+9°26′	−2°20′	−13°23′	−21°27′	−23°15′
30	−17°43′		+3°41′	+14°42′	+21°44′	+23°11′	+18°34′	+9°05′	−2°43′	−13°43′	−21°37′	−23°11′
31	−17°26′		+4°05′		+21°53′		+18°19′	+8°43′		−14°03′		−23°07′

TABLE 5

Hour-angle of the sun at the instant of sunrise for the summer (northern hemisphere) or winter (southern hemisphere) solstices

LATITUDE	HOUR-ANGLE	LATITUDE	HOUR-ANGLE	LATITUDE	HOUR-ANGLE
0	6h 00m	32	7h 03m	48	7h 56m
2	6h 04m	33	7h 06m	49	8h 00m
4	6h 07m	34	7h 09m	50	8h 05m
6	6h 11m	35	7h 11m	51	8h 10m
8	6h 14m	36	7h 14m	52	8h 16m
10	6h 18m	37	7h 17m	53	8h 21m
12	6h 22m	38	7h 19m	54	8h 27m
14	6h 25m	39	7h 23m	55	8h 34m
16	6h 29m	40	7h 26m	56	8h 41m
18	6h 33m	41	7h 29m	57	8h 48m
20	6h 37m	42	7h 33m	58	8h 57m
22	6h 40m	43	7h 36m	59	9h 06m
24	6h 45m	44	7h 40m	60	9h 16m
26	6h 49m	45	7h 43m	61	9h 27m
28	6h 54m	46	7h 47m	62	9h 40m
30	6h 59m	47	7h 51m	63	9h 55m
31	7h 01m				

The above table gives the true time of sunset for the summer solstice, the longest day of the year in the northern hemisphere. To determine the time of sunrise, subtract the hour-angle, HA, from 12 hours; thus, for a given latitude, the limits to be marked on a dial may be calculated.

In the southern hemisphere the readings are identical, but apply to the winter solstice.

To calculate the length of the shortest day of the year (winter solstice) subtract the appropriate reading above from 12 hours.

So for latitude 45° the length of the shortest day of the year will be

$$2 \times (12 - HA) = 2 \times (12 - 7h\ 43m) = 8h\ 34m.$$

References

BEDOS DE CELLE *La Gnomonique pratique*, Paris, 1760

BENCKER, H. *Étude descriptive du gnomon d'Annecy* (*La Revue savoisienne*, no. 2, 1963)

BIGOURDAN, G. *Gnomonique*, Paris, 1956

BOURSIER, CH. *Huit cents devises de cadrans solaires*, Paris, 1936

DRECKER, J. *Gnomone und Sonnenuhren*, Aix-la-Chapelle, 1909 *Ephémérides nautiques*

FRIOCOURT, G. *Tables de logarithmes et de navigation*, Paris, 1925

KÜHNELT, PROF DR HARRO H. *Die Sonnenuhren in Nordtirol* (Tiroler Heimat, 1952, 1954)

KÜHNHANSS, H. *Docet umbra* (Alte und neue Kunst, Zürich) *La Connaissance des Temps*

LIVET, C. S. F. *Gnomonique*, Metz, 1839

LOSCHNER, DR HANS *Über Sonnenuhren*, Graz, 1905

LOSKE, L. M. *Die Sonnenuhren*, Berlin, 1958

MARCHAND, F. *Le cadran de Brou* (Extrait des Annales de la Société d'Émulation et d'Agriculture de l'Ain)

MASSENET and HARDANT *Traité d'astronomie nautique*, Paris, 1921

MASSENET and HARDANT *Traité de navigation*, Paris, 1923

MAYALL, R. N. and M. *Sundials*, Boston, 1962

MICHEL, HENRI *Catalogues des cadrans solaires du Musée de la Vie wallonne*, Liège, 1953

MICHEL, HENRI *Méthodes astronomiques des hautes époques chinoises* (Les Conférences du Palais de la Découverte)

MOUREAU, C. *Lire correctement le cadran solaire*, Carcassonne, 1962

NOEL, PIERRE *Construction des cadrans solaires*

PERRIN (AMIRAL) *Nouvelles tables de calculs nautiques*

POHL, HELGA *L'Homme à la poursuite du temps*, Paris, 1957

REY-PAILHARDE, A. ROME and AUG. FAVOT *Le cadran solaire à réflexion du Lycée de Jeunes Filles de Grenoble* (Bulletin de la Société scientifique de l'Isère, tome 42)

RICHER *La Gnomonique universelle*, Paris, 1701

RIGHINI *La Tradizione Astronomica Fiorentina e l'Osservatorio di Arceti* (Rivista di Storia della Scienza, vol IV, fasc 2, 1962)

RIVARD *La Gnomonique*, Paris, 1767

SAINTE-MARIE-MADELEINE (DOM PIERRE DE) *Traité d'horlogiographie*, 1681

SHEPHARD, DR G. C. *Queens College Dial*, Cambridge

Plates

*NOTE: in this edition the colour plates A–C appear in black and white
in the text and in colour on the inside back cover.

Index